FAR OUT MATH!

SCIENCE WITH SIMPLE THINGS SERIES

Conceived and written by **RON MARSON**

Illustrated by **PEG MARSON**

TOPS LEARNING SYSTEMS

342 S Plumas Street
Willows, CA 95988

www.topscience.org

Copyright © 2003 by TOPS Learning Systems. All rights reserved.
No part of this book may be reproduced in any manner whatsoever without written permission from the publisher, except as explicitly stated below:

NASA, which funded this project (thank you, NASA!) has unrestricted reproduction and dissemination rights. Individual owners of this book have our permission to make photocopies for personal classroom or homeschool use. Reproduction of these pages by schools, school systems or teacher training programs for wider dissemination, or by anyone for commercial sale, is strictly prohibited unless licensing for such has been purchased from TOPS Learning Systems.

ISBN 978-0-94-100843-3

Teach abstract, sometimes challenging standards-based concepts in science and math in creatively concrete ways. Students study orders of magnitude, exponents, logarithms and other fundamental concepts that enable scientists to launch observatories like GLAST into space, and to interpret the data received.

Table of Contents

		PAGE
	Introduction	4 – 9
	welcome / standards	4
	history / overview	5
	getting ready	6
	tracking sheet	7
	test questions / answers	8 – 9

A *Sliding Scales*

	PAGE
A1: Adding Slide Rule	10 – 14
plastic hairlines	15
adding slide rule	16
Glastoids	17
A2: Multiplying Slide Rule	18 – 22
multiplying slide rule	23

B *Log Tape*

	PAGE
B1: Log Tape	24 – 27
log tape	31
B2: String Calculator	28 – 30
B3: No Strings Calculator	32 – 33

C *Base-Two Slide Rule*

	PAGE
C1: Base-Two Slide Rule	34 – 37
base-two slide rule	43
C2: Exponential Ups and Downs	38 – 39
C3: Log Algebra	40 – 42

D *Classic Log Scales*

	PAGE
D1: Log Ruler	44 – 45
log ruler	46
playing scales	47
D2: Scientific Notation	48 – 50
D3: Classic Slide Rule	52 – 55
classic slide rule	51

E *Log Graphs*

	PAGE
E1: Slide-Rule Graph	56 – 57
slide rule graph	62
E2: Inverse Functions	58 – 59
inverse function graph	63
E3: Ironing Out Curves	60 – 61
linear/log graph	64

Welcome, Dear Educator, ...

... to **FAR OUT MATH!** This book is the first of 3 activity units (published one per year) designed to inform students about NASA's 2006 space shot called GLAST.

It's not hard to say what GLAST is, or to outline what it will do. GLAST stands for the Gamma-ray Large Area Space Telescope. It looks like a "flying" box with solar panel "wings." GLAST is designed to absorb, track and measure gamma ray photons across enormous energy ranges. It will open a window of observation and understanding into our Universe at the highest end of the electromagnetic spectrum, enabling us to better understand the dance of matter and energy at Nature's most extreme and energetic levels.

It's more difficult to bring GLAST down to earth, into the imagination and grasp of high school math and science students. How can TOPS create high-interest, hands-on activities that inform students about the GLAST mission, **and** teach to the standards in meaningful ways?

For our first year, we decided to launch into logarithms, an important mathematical tool that has helped us design space telescopes, and make sense of the huge ranges of data received back on earth. Those of us who are baby-boomers (or older) remember calculating with logarithms in high school, and carrying slide rules into our physics classes. Would our current generation of students, raised on scientific calculators, be well served and meaningfully educated by bringing back this bit of not-so-ancient history?

Do **you** really understand logarithms? Did you learn them in high school, then have to relearn them in college, then learn them again to teach second year algebra? Our experience at TOPS is that logarithms have been eminently forgettable. But not any longer! After extensive development and testing, we think we've come up with an accessible, creative program that you and your students won't easily forget, and can actually enjoy.

Whether you think we've achieved mission impossible (informing about GLAST and teaching stuff that future rocket scientists need to know), or that our first attempt is a fizzle, we'd love to hear from you.

Please direct your comments and suggestions to TOPS at the address on our copyright page, and send a copy to Lynn Cominsky, our program director.

Sincerely,

Ron Marson

FAR OUT MATH! ties in with **National Math and Science Standards**
in the following areas. See page 6 for content details.

Math:

Understand and compare the properties of classes of functions, including exponential, polynomial, rational, logarithmic, and periodic functions.... Appreciate that seemingly different mathematical systems may be essentially the same.... Make decisions about units and scales that are appropriate for problem situations involving measurement.... Develop a deeper understanding of very large and very small numbers and of various representations of them.

(National Council of Teachers of Mathematics)

Science:

Develop abilities necessary to do scientific inquiry.... A variety of technologies, such as hand tools, measuring instruments and calculators, should be an integral component of scientific investigations.... Mathematics plays an essential role in all aspects of an inquiry. For example, measurement is used for posing questions, formulas are used for developing explanations, and charts and graphs are used for communicating results.

(The National Science Education Standards)

Overview / A Brief History Of Logarithms

In the early 1600's John Napier, Henry Briggs, and other mathematicians worked out a method to express positive numbers as powers of ten:

$3 = 10^{0.47712}$ $5 = 10^{0.69897}$ $15 = 10^{1.17609}$

They called these base-10 exponents "logarithms", an invented word suggesting "logical arithmetic." Briggs published these logarithms in a log table. Here's a simplified example:

log 1 = 0.00000	... log 13 = 1.11394
log 2 = 0.30103	log 14 = 1.14613
log 3 = 0.47712	log 15 = 1.17609
log 4 = 0.60206	log 16 = 1.20412
log 5 = 0.69897 ...	log 17 = 1.23045 ...

Any good algebra students knows that numbers in the same base are multiplied by adding their exponents, and divided by subtracting. Using Briggs' table, it was a straightforward process to **(1)** look up the exponents (logs) of numbers one wished to multiply or divide, **(2)** add or subtract these logs, then **(3)** use the table in reverse to find the product or quotient (the antilog):

To MULTIPLY:	*To DIVIDE:*
3 x 5 = ?	15 / 5 = ?
log 3 = 0.47712	log 15 = 1.17609
+ log 5 = 0.69897	- log 5 = 0.69897
1.17609	0.47712
antilog 1.17609 = 15	antilog 0.47712 = 3
Thus, 3 x 5 = 15	*Thus, 15 / 5 = 3*

This may seem like a complicated way to do simple arithmetic. But if the numbers were not so simple, and your only alternative were to multiply and divide huge numbers longhand, then you might prefer to add or subtract their logs instead.

Soon after Briggs published his log tables, Edmund Gunter invented the logarithmic scale, so constructed that the position of each number on the line was in direct proportion to the value of its log (exponent), as illustrated on the scale below:

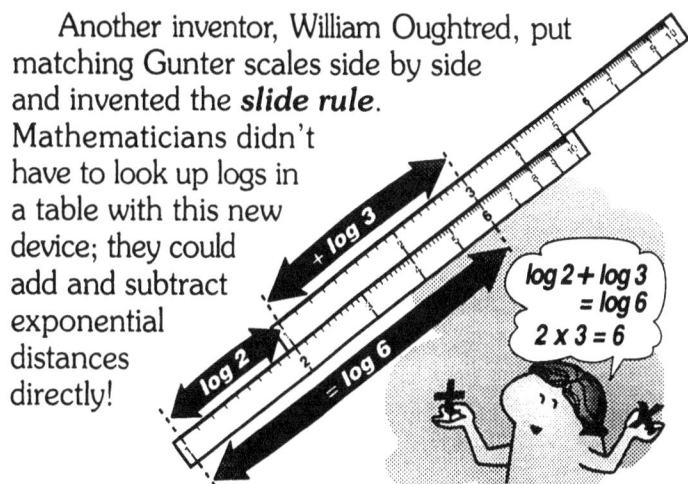

Another inventor, William Oughtred, put matching Gunter scales side by side and invented the **slide rule**. Mathematicians didn't have to look up logs in a table with this new device; they could add and subtract exponential distances directly!

Slide rules, for Western scientists and engineers, remained the calculator of choice for the next three centuries. But in the mid-1970's, mass production of new, hand-held digital calculators displaced the slide rule.

Today, we seldom use logarithm tables, because with a calculator, any arithmetic operation is just about as easy as any other. (Actually, a little guy inside your calculator does use logarithms, every time you push the button to take a power or a root. The process is just hidden from view!)

But logarithms are much more than an historical oddity. They are certainly relevant to GLAST astronomers: as an essential tool for solving exponential equations; as a way to compress huge ranges of gamma-ray energies into exponential orders of magnitude; as a tool to transform curving power equations into straight line graphs for simplified analysis.

And logs are a working part of our everyday lives! We measure earthquakes on a base-10 Richter scale. Music frequencies double octave by octave up the scale in base-2 logarithms. The decibel scale for loudness and the magnitude scale for star brightness are both logarithmic, devised to mimic the way we actually hear and see!

Well, then, let's learn how logs work and see what they can do. Welcome to **FAR OUT MATH!**

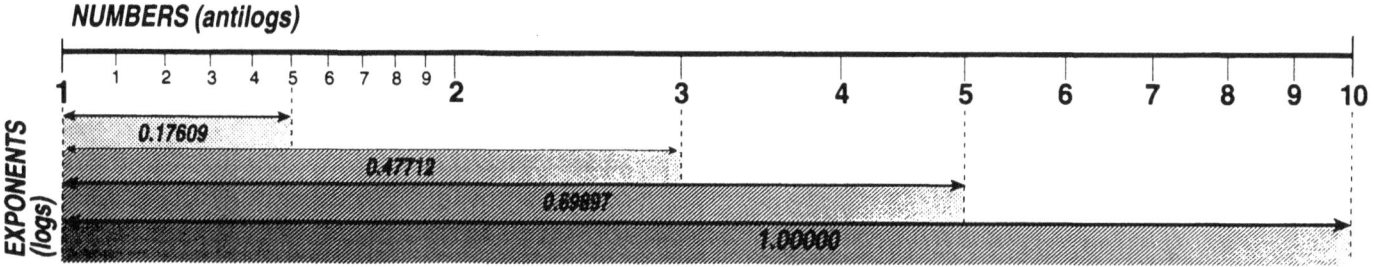

Getting Ready

1. Review the scope and sequence outlined below, then page through the book to fill in conceptual details.

Lesson / Time		Main Idea(s)
A1	1h:15m	Slide rules, calibrated in a linear scale of numbers, add and subtract distance.
A2	1h:25m	Slide rules, calibrated in a logarithmic scale of exponents, multiply and divide distance.
B1	1h:30m	Logs are exponents *(log N = exp)*. Antilogs are numbers *(antilog exp = N = 10^{exp})*.
B2	1h:15m	Logarithm laws reduce the complexity of calculating ordinary numbers by 1 notch.
B3	:20m	To calculate with logarithms: *(1)* take logs; *(2)* apply laws; *(3)* take the antilog.
C1	:50m	All the log relationships you learned in base-10 also apply in base-2.
C2	:20m	Translate antilog equations into log equations by "pulling down the exponent."
C3	:40m	Solve exponential equations by applying base-ten or base-two logarithms.
D1	1h:0m	Significant figures are all figures you know for sure, plus the last estimated digit.
D2	1h:10m	Translate logs and antilogs using scientific notation as an intermediate step.
D3	1h:30m	A classic slide rule calculates the numbers, but *you* figure the decimal place.
E1	:40m	Graph N *vs* log N to generate a logarithmic scale. Convert it to a slide rule.
E2	:50m	Exponential equations and log equations are inverse functions.
E3	:50m	Log-log graphs transform higher order equations into straight lines.

2. Choose a lesson sequence that fits your time frame and curriculum goals. **STRAND A** lessons offer a gentle start that keeps all options open. Or, start with **STRAND B** to cut to the conceptual foundations of logarithms a bit faster. The *Log Tape* constructed in lesson *B1* is an important reference used in many lessons that follow. **STRAND C** reinforces the common logarithms presented in **STRAND B** from a base-2 perspective. **STRAND D** broadens into work with significant figures, scientific notation, and estimating the decimal place. **STRAND D** can stand alone, but the learning curve is steeper: it requires students to operate the *Classic Slide Rule* without having used earlier practice versions. **STRAND E** also stands alone as a 3-lesson unit on graphing, assuming your students already have a fundamental working knowledge of logarithms.

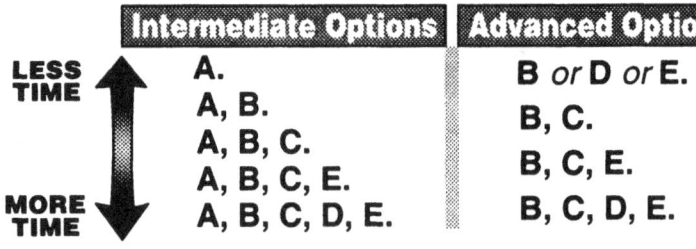

	Intermediate Options	Advanced Options
LESS TIME	A.	B *or* D *or* E.
	A, B.	B, C.
	A, B, C.	B, C, E.
	A, B, C, E.	B, C, E.
MORE TIME	A, B, C, D, E.	B, C, D, E.

3. If you wish to use the optional *Tracking Sheet* for students (opposite), cross out any lessons you plan to skip.

4. Cut freezer bags to size and gather other materials. See teaching notes 11 for our simple **materials list**.

5. Photocopy student materials for your first few activities. If you copy on both sides of the paper, only pages for lessons *A1* and *A2* need to be stapled together.

6. Jumbo-clip sets of duplicates, and organize a resource shelf or table. Students will each need a manila folder (or equivalent) to store their *Tracking Sheet*, manipulatives, and completed lessons.

7. Direct your students to pick up a *Tracking Sheet* and a *Plastic Hairlines* supplement (hairlines are not needed if you will use strand B or E only), and begin work.

Suggested Teaching Strategies

✱ Allow students to move through each lesson at their own pace.

✱ Require check-points. Write your initials in each box (see opposite page), before starting a new lesson.

✱ Interrupt individualized activity at any point to present optional introductory materials from the lesson notes, or to clarify points of general interest.

✱ Post answer keys for self checking.

✱ Communicate behavior expectations.

✱ Explain grading procedures. Use any combination of the review/test questions (pp 8-9), plus questions of your own, to determine how well students have mastered key concepts.

Tracking Sheet

1. Tape shut one edge of a manila folder as shown to make a "pocket." Write your name on the tab.
2. Get the **Plastic Hairlines** supplement (page 15). Follow the directions, using the materials provided.
3. Continue with the lesson box your teacher wants you to complete first. All the pages you need for each lesson are listed inside its box.
4. As you finish each lesson, ask your teacher to check your work and initial the corresponding box. Do this *before* gathering new activity pages specified in the next lesson box.

USAGE KEY:
Activity Sheet(s) (page #'s)
supplement (page #'s)
(previous supplement)
TIME = hours :minutes
teacher check ☐

	1	2	3
A	**Adding Slide Rule** (pp 10, 12, 14) adding slide rule (p 16) glastoids (p 17) TIME = 1 h :15 m *checkpoint* ☐	**Multiplying Slide Rule** (pp 18, 20, 22) multiplying slide rule (p 23) TIME = 1 h :25 m *checkpoint* ☐	
B	**Log Tape** (pp 24, 26) log tape (p 31) TIME = 1 h :30 m *checkpoint* ☐	**String Calculator** (pp 28, 30) (log tape) TIME = 1 h :15 m *checkpoint* ☐	**No Strings Calculator** (p 32) (log tape) TIME = :20 m *checkpoint* ☐
C	**Base-Two Slide Rule** (pp 34, 36) base-two slide rule (p 43) TIME = :50 m *checkpoint* ☐	**Exponential Ups & Downs** (p 38) (base-2 slide rule) (log tape) TIME = :20 m *checkpoint* ☐	**Log Algebra** (pp 40, 42) (log tape) (base-2 slide rule) TIME = :40 m *checkpoint* ☐
D	**Log Ruler** (p 44) log ruler (p 46) playing scales (p 47) TIME = 1 h :0 m *checkpoint* ☐	**Scientific Notation Lubes Logs** (pp 48, 50) (log ruler) TIME = 1 h :10 m *checkpoint* ☐	**Classic Slide Rule** (pp 52, 54) classic slide rule (p 51) TIME = 1 h :30 m *checkpoint* ☐
E	**Slide-Rule Graph** (p 56) slide rule graph (p 62) (log tape) TIME = :40 m *checkpoint* ☐	**Inverse Functions** (p 58) inverse functions graph (p 63) (base-2 slide rule) (log tape) TIME = :50 m *checkpoint* ☐	**Ironing Out Curves** (p 60) linear/log grids (p 64) TIME = :50 m *checkpoint* ☐

Review / Test Questions

Photocopy these test questions. Cut out those you wish to use, and tape them onto white paper. Include questions of your own design, as well. Crowd them all onto a single page for students to answer on their own papers, or leave space for student responses after each question, as you wish. Duplicate a class set, and your custom-made test is ready to use. Use leftover questions as a class review in preparation for the final exam.

activity A1, A2
a. Explain how you would use a pair of rulers to show that $3 + 6 = 9$. Provide a clearly labeled drawing to illustrate your answer.

b. What if the figures on your drawing represented base-10 exponents (logs) instead of ordinary numbers? Rewrite the addition problem in part a, showing how it becomes a multiplication problem.

activity A2
Slide rules add and subtract distances to multiply and divide numbers. How is this possible?

activity B1
Use your *Log Tape* to fill in the missing numbers:

a. $2 = 10^{\text{-----}}$ = antilog _____.
Thus, log _____ = _____.

b. $20 = 10^{\text{-----}}$ = antilog _____.
Thus, log _____ = _____.

activity B1
Use your *Log Tape* to fill in the missing numbers:

a. $3 = 10^{\text{-----}}$ = antilog _____.
Thus, log _____ = _____.

b. $.03 = 10^{\text{-----}}$ = antilog _____.
Thus, log _____ = _____.

activity B1
The log of **a number** is _____.
The antilog of _____ is _____.

activity B2, B3
a. Fill in the missing number:
$\log 5 + \log 2 = \log$ _____.

b. Use your *Log Tape* to evaluate each log, and prove the equality.

activity B2, B3
a. Fill in the missing number:
$\log 5 - \log 2 = \log$ _____.

b. Use your *Log Tape* to evaluate each log, and prove the equality.

activity B2, B3
a. Fill in the missing number:
$3 \log 3 = \log$ _____.

b. Use a *Log Tape* to evaluate each log, and prove the equality.

activity B2, B3
a. Fill in the missing number:
$(\log 25) + 2 = \log$ _____.

b. Use your *Log Tape* to evaluate each log, and prove the equality.

activity C1, C2
The antilog$_2$ of **an exponent** is _____.
The log$_2$ of _____ is _____.

activity C1, C2
Using your *Base-2 Slide Rule* as a reference, complete these equivalent equations:

$\log_2 64 + \log_2 2 =$ $\log_2$ _____	___ × ___ = ___
	$\dfrac{4{,}096}{512} =$ ___
$5 \log_2 4 =$ $\log_2$ _____	
	$\sqrt[3]{2{,}097{,}152} =$

activity C1, C2
Complete this table:

$2^{13} =$		
	$\log_2 128 =$	
		antilog$_2$ 0 =
$10^{-3} =$		
	$\log 70 =$	
		antilog 2.845 =

activity C3
Solve for x. Show your work: $4^x = 15$

activity C3
Solve for x in base-2, then again in base-10. Show your work: $8^x = 32$

activity D1
Read your *Log Ruler* as accurately as possible.
$25 = 10^{\text{-----}}$. Thus, log 25 = _____.
$250 = 10^{\text{-----}}$. Thus, log 250 = _____.
$10^{0.500} =$ _____. Thus, antilog 0.500 = _____.
$10^{\overline{2}.500} =$ _____. Thus, antilog $\overline{2}.500 =$ _____.

activity D2
a. Complete this table of logs and antilogs:

antilog	scientific notation	log
514		
22.8		
		4.069

b. Solve using logs. Show your work.
$514 \times 22.8 = ?$

activity D3
Solve these problems on your slide rule. Report significant figures.

a. $1.234 \times 12.34 = ?$

b. $2580. + 1850. = ?$

activity D3
Your *Classic Slide Rule* uses log distances to multiply and divide numbers, even though no logs are printed on it anywhere! How is this possible?

activity E1, E2
a. Label equations *I*, *II*, and *III* in the boxes provided.

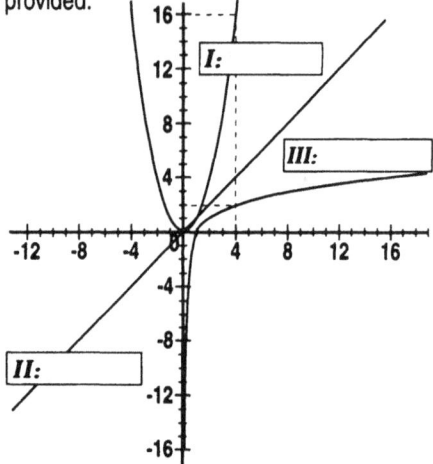

b. State the *inverse* of...
equation *I*: _____
equation *III*: _____

activity E3
a. Label equations *I*, *II*, and *III* for this log-log grid in the boxes provided.

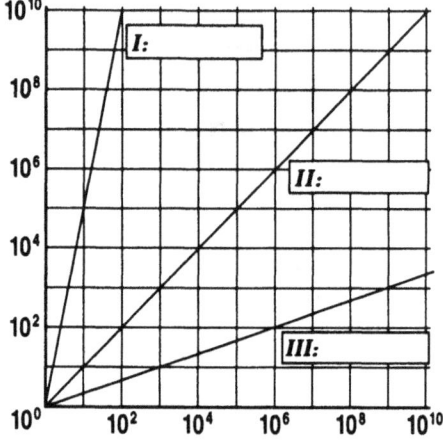

b. State the inverse of...
equation *I*: _____
equation *III*: _____

Answers

activity A1, A2

a. Slide the rulers relative to each other, so a distance of 3 units on the bottom plus a distance of 6 units on the top add up to a distance of 9 units, (where 6 meets 9).

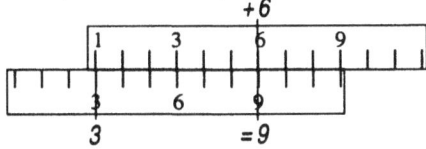

b. $10^3 \times 10^6 = 10^9$ (or)
(1,000) (1,000,000) = 1,000,000,000 (or)
(1 thousand)(1 million) = 1 billion

activity A2

Slide rules add and subtract distances that are calibrated in base-10 exponents, not ordinary numbers. These exponents (logs) multiply base-10 numbers when added together ($10^a \times 10^b = 10^{a+b}$), and divide base-10 numbers when subtracted ($10^a \div 10^b = 10^{a-b}$).

activity B1

a. $2 = 10^{.301} = $ antilog .301
Thus, log 2 = .301

b. $20 = 10^{1.301} = $ antilog 1.301
Thus, log 20 = 1.301

activity B1

a. $3 = 10^{.477} = $ antilog .477
Thus, log 2 = .477

b. $.03 = 10^{\overline{2}.477} = $ antilog $\overline{2}.477$
Thus, log .03 = $\overline{2}.477$

activity B1

The log of **a number** is **an exponent**.
The antilog of **an exponent** is **a number**.

activity B2, B3

a. log 5 + log 2 = log 10
b. .699 + .301 = 1.000

activity B2, B3

a. log 5 − log 2 = log 5/2 = log **2.5**
b. .699 − .301 = **.398**

activity B2, B3

a. 3 log 3 = log 3^3 = log **27**
b. 3 (.477) = **1.431**

activity B2, B3

a. (log 25) ÷ 2 = log $\sqrt{25}$ = log **5**
b. 1.398 ÷ 2 = **0.699**

activity C1, C2

The antilog$_2$ of **an exponent** is **a number**.
The log$_2$ of **a number** is **an exponent**.

activity C1, C2

Equivalent equations:

log$_2$ 64 + log$_2$ 2 = log$_2$ **128**	64 × 2 = **128**
log$_2$ 4096 − log$_2$ 512 = log$_2$ **8**	$\frac{4,096}{512}$ = **8**
5 log$_2$ 4 = log$_2$ **1024**	4^5 = **1024**
$\frac{\log 2,097,152}{3}$ = log **128**	$\sqrt[3]{2,097,152}$ = **128**

activity C1, C2

Equivalent equations:

2^{13} = 8192	log$_2$ 8192 = 13	antilog$_2$ 13 = 8192
2^7 = 128	log$_2$ 128 = 7	antilog$_2$ 7 = 128
2^0 = 1	log$_2$ 1 = 0	antilog$_2$ 0 = 1
10^{-3} = .001	log .001 = −3	antilog −3 = .001
$10^{1.845}$ = 70	log 70 = 1.845	antilog 1.845 = 70
$10^{2.845}$ = 700	log 700 = 2.845	antilog 2.845 = 700

activity C3

$4^x = 15$
log 4^x = log 15
x log 4 = log 15
x = log 15 / log 4
x = 1.176 / .602 = **1.953**

activity C3

Base-2:   **Base-10:**
$8^x = 32$ $8^x = 32$
log$_2 8^x$ = log$_2$ 32   log 8^x = log 32
x log$_2$ 8 = log$_2$ 32   x log 8 = log 32
x = log$_2$ 32 / log$_2$ 8 x = log 32 / log 8
x = 5 / 3 = **1.667**   x = 1.505 / .903 = **1.667**

activity D1

25 = $10^{1.398}$ Thus, log 25 = **1.398**
250 = $10^{2.398}$ Thus, log 250 = **2.398**
$10^{0.500}$ = 3.162 Thus, antilog 0.500 = **3.162**
$10^{2.500}$ = 316.2 Thus, antilog 2.500 = **316.3**

activity D2

a.

antilog	scientific notation	log
514	5.14 × 10^2	2.711
22.8	2.28 × 10^1	1.358
11720	1.172 × 10^4	4.069

b.   514 × 22.8 = ?
2.711 + 1.358 = 4.069
antilog 4.068 = 11720
Thus, 514 × 22.8 ≈ 11720

activity D3

a. 1.234 × 12.34 = **15.25**
b. 2580 ÷ 1850 = **1.395**

activity D3

A **Classic Slide Rule** is a pair of base-10 exponent scales, labeled not with exponents, but instead with numbers (antilogs) positioned to correspond to the values of their exponents. When the slide rule adds and subtracts distances between numbers, it is really adding and subtracting their logs, and thus multiplying and dividing the corresponding numbers.

activity E1, E2

a.

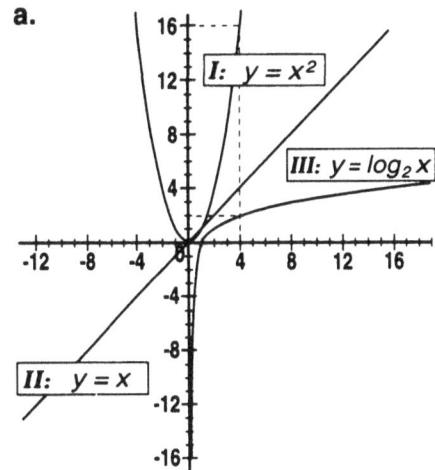

b. Inverse of equation I: $y = \sqrt[2]{x}$
Inverse of equation III: $y = 2^x$

activity E3

a.

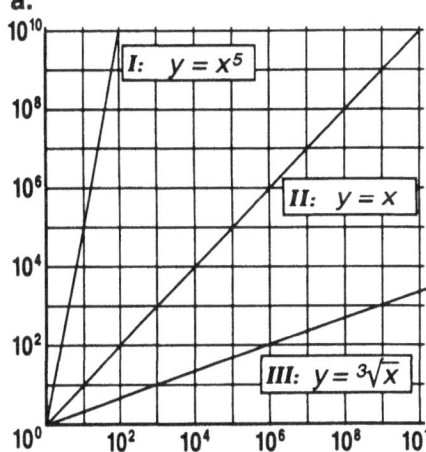

b. Inverse of equation I: $y = \sqrt[5]{x}$
Inverse of equation III: $y = x^3$

ADDING SLIDE RULE

Activity A1
1 of 3 pages

1. Get the *Plastic Hairlines* supplement and follow the directions. Then return here.

2. Get the *Adding Slide Rule* supplement and follow the directions. Return here again.

3. Fit the folded C scale completely *inside* (not over) the folded D scale, so both scales meet as shown. (If you did this right, you'll see a name box on the back.)

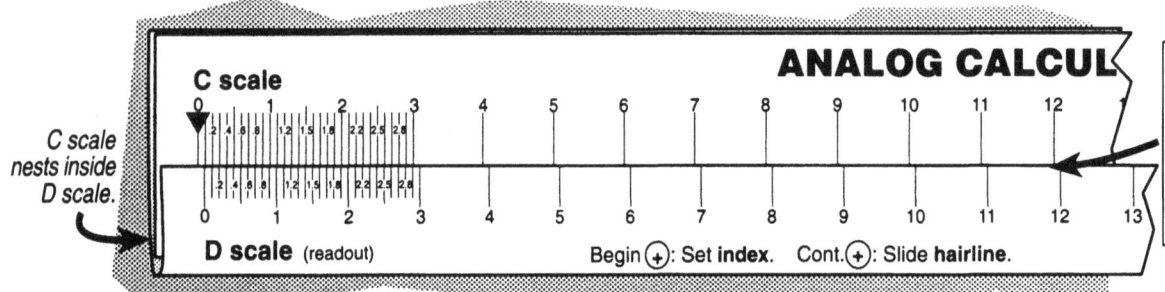

4. Write your name in the name box.

5. Wrap one of your hairlines across the name box (inked-side down), so the ink line *perfectly* overlaps. Pull firmly together and tape. (Don't tape the paper!)

6. Check that the hairline fits snugly, yet still moves freely. (If too tight or too loose, peel the tape up and try again.)

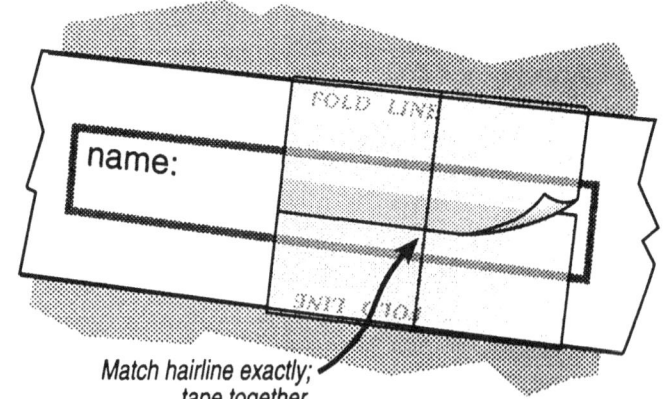

TEACHER CHECK:

7. Identify and label these slide rule parts on the dotted lines:
Hairline • *C scale* • *Index* • *D scale*

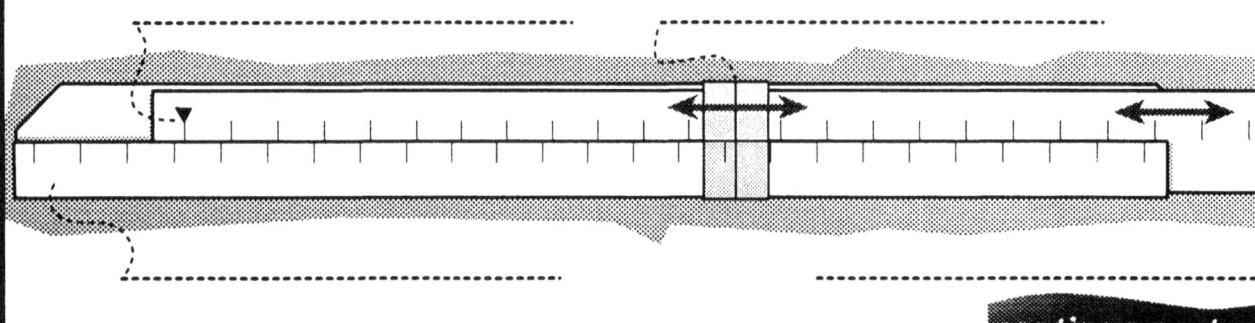

continue next page

OVERVIEW / OBJECTIVES

Students will construct *Adding Slide Rules*, scaled with linear calibrations like ordinary rulers. They will learn to move these scales (and the hairline) relative to each other in ways that add and subtract distance, thus calculating sums and differences.

TIME: 1 hour and 15 minutes.

INTRODUCTION

★ Use this (and all), introductions as *optional* overviews. If you choose to present this one, please keep in mind that you are *not* trying to teach your students how to operate a slide rule. (They will pick this up quite naturally as they interact with these TOPS lessons.) Rather, your are demonstrating how sliding rulers add and subtract distance.

Snip sections of masking tape (5-10 cm long); roll them sticky-side-out. Use these to fix a meter stick to your board or classroom wall. Rest a second meter stick above the first. The bottom meter is *fixed*; the top meter *moveable*.

Slide the zero calibration on the moveable meter 20 cm to the right of the fixed meter. Use your hand as an imaginary hairline to show how numbers on the top, plus 20, always equal numbers on the bottom.

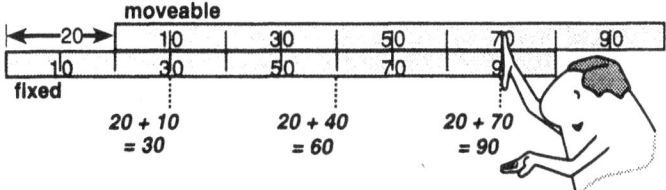

Slide the top meter stick other distances (30 cm, 40 cm...) relative to the bottom to demonstrate other additives.

Then demonstrate the inverse process of subtraction. Start with your "hairline" on any fixed-scale distance, say 90 as illustrated below. Notice that the movable-scale distance, 70, subtracts to leave a difference of 20, indicated under the zero calibration. Notice how 20 is the common difference for *any* number pair related by the hairline.

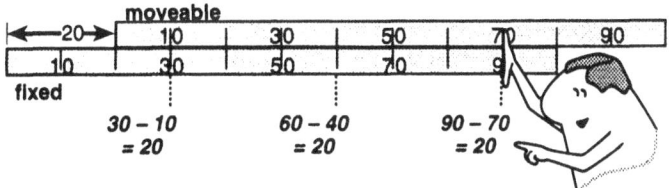

Slide the top meter stick other distances relative to the bottom to demonstrate other subtraction differences.

★ Make an *Adding Slide Rule* yourself to show your students what they will be making. This will help orient them to how the *Plastic Hairlines* are used, and how the scales fit together. This will also give you, the teacher, an appreciation for the clarity and integrity of the directions you'll be asking students to follow. When they come to you for help, ask them to read the direction or problem that puzzles them aloud. This is often enough to clear confusion. Remember, our goal as educators is to foster learning independence.

NOTES (this page)

1-2. Point out that steps 1 and 2 are giant steps. Each involves the completion of a separate, supplementary sheet of instructions before students continue with step 3.

☞ *See Plastic Hairlines supplement, pg 15.*

2-7. The double layer of plastic is folded *inside* the paper, as illustrated, not outside. The tape and paper clips serve to hold this plastic "sandwich" firmly together during trimming and cutting. Mature students who recognize that these layers must not shift may accomplish this task with a firm, careful grip, eliminating the need for clipping or taping altogether.

☞ *See Adding Slide Rule supplement, pg 16.*

2. Demonstrate how to accurately pinch-fold a dotted line between thumb and finger along its *entire* length. When this is done accurately, the dots (or light dashes) don't end up on one side of the fold or the other, but precisely *on* the fold, where they belong. Caution against getting the fold started, then creasing it against a table top for the remainder of the line; usually the dashed line will stray to one side or the other.

Return to this page. ☜

6. Inspect the slide rule edges, to check that all dotted lines remain *on* the folds. Direct students to refold (by pinching between thumb and finger) as necessary. Good folds make the slide rule a pleasure to use. Bad folds are reversible.

ANSWERS (this page)

7. | Index | Hairline |
    | D scale | C scale |

MATERIALS

☐ A double layered rectangle of plastic cut from a transparent storage bag. Sandwich bags work, but thicker plastic cut from freezer bags has better handling qualities. Larger bags may be divided into halves or quarters for conservation. Each double layer should measure **8 cm x 13.5 cm** or larger, with at least 1 bonded edge holding both layers together. See page 15 for the paper template students will use with this plastic.

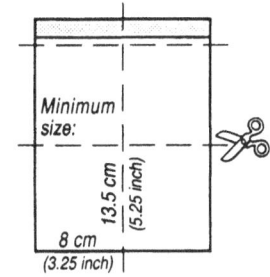

☐ Scissors that are sharp enough to cut through plastic bags with minimum drag.

☐ Paper clips.

☐ Tape. Masking tape and clear tape both work fine.

☐ A ballpoint pen that writes on plastic.

☐ Copier or notebook paper. Or recycle scrap paper printed on one side for conservation.

☐ No other materials will be needed for this unit except **manila folders** *(see page 7)*, and **string** *(see notes 29)*. Students will occasionally use **calculators**, and require **rulers** (any straightedge) in STRAND E. You may also need 2 **meter sticks** for optional lesson introductions.

NOTES: **Activity A1**

ADDING SLIDE RULE

8. Try this **addition** problem: $\overset{I}{\underset{D}{8}} + \overset{H}{\underset{C}{5}} = \overset{H}{\underset{D}{13}}$

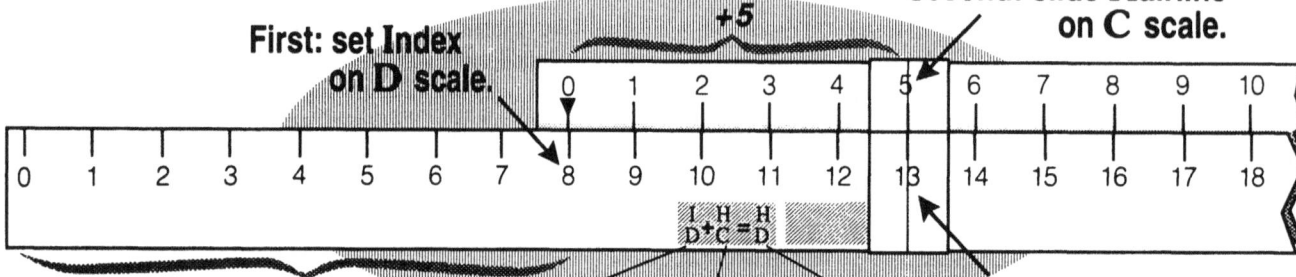

First: set Index on D scale.
Second: slide Hairline on C scale.
Third: read Hairline on D scale.

9. Find the grey "+" box on your slide rule. Summarize how the letters help you **add**: 8 + 5 = 13

First ...	Second...	Third...

10. Find these **sums** on your slide rule:

$\overset{I}{\underset{D}{4}} + \overset{H}{\underset{C}{4}} =$ ____ $\underset{D}{H}$	$\overset{I}{\underset{D}{1}} + \overset{H}{\underset{C}{21}} =$ ____ $\underset{D}{H}$	$\overset{I}{\underset{D}{10}} + \overset{H}{\underset{C}{4}} =$ ____ $\underset{D}{H}$	$\overset{I}{\underset{D}{1.5}} + \overset{H}{\underset{C}{1.5}} =$ ____ $\underset{D}{H}$
6 + 9 = ____	21 + 1 = ____	10 + 2 = ____	1.1 + 1.5 = ____
13 + 11 = ____	8 + 11 = ____	10 + 0 = ____	0.1 + 0.2 = ____

11. "D" is the distance from 0 to 1 on your slide rule. Complete each equation in terms of this **unit distance**.

(0 to 1) = 1 D
(0 to ___) = 4 D
(___ to 12) = 12 D
(0 to 18) = ___ D
(0 to ___) = ___ D

a. The size (magnitude) of each number on this slide rule is proportional to its distance from what particular number?

_____

b. What physical variable does this slide rule keep track of?

_____

12. Add 7 + 12 on your slide rule. How does it "calculate" an answer of 19?
(Hint: Write about units of distance.)

13. Add calibrations (marks) to each end of your slide rule, so you can solve these problems:

13 + 13 = ____   -1 + 5 = ____   -1 + 26 = ____

a. What new marks did you add?

b. Invent a problem that exceeds the capacity of your slide rule.

continue next page

NOTES (this page)

8-9. The two grey boxes, printed on all slide rules, neatly summarize *what* to move (top symbol), *where* to move it (bottom symbol), and *when* to move it (in what order). Notice that one box describes an *addition* problem, and the other shows *subtraction*.

10. Notice that these addition problems always *start* by moving the index, and *continue* by moving the hairline.

11. Slide rules are called "analog computers" because they operate with numbers that are analogous to direct physical measurement, in this case distance. Modern "digital computers," by contrast, operate with numbers expressed directly as digits (ones and zeros).

ANSWERS (this page)

9. First... *Set the Index over "8" on the D scale.*
Second... *Slide the Hairline to "5" on the C scale.*
Third... *Read "13" under the Hairline on the D scale.*

10.
4 + 4 = *8*	1 + 21 = *22*	10 + 4 = *14*	1.5 + 1.5 = *3.0*
6 + 9 = *15*	21 + 1 = *22*	10 + 2 = *12*	1.1 + 1.5 = *2.6*
13 + 11 = *24*	8 + 11 = *19*	10 + 0 = *10*	0.1 + 0.2 = *0.3*

11. (0 to 1) = 1 D
(0 to *4*) = 4 D
(*0* to 12) = 12 D
(0 to 18) = *18* D
(0 to __) = __ D *(blanks have same number.)*

11a. The size (magnitude) of each number on this slide rule is proportional to its distance from *zero* (the index).

11b. This slide rule (an analog computer) keeps track of *distance*.

12. *The slide rule "calculates" that 7 + 12 = 19 by adding 7 units of distance to 12 units of distance to get a total of 19 units of distance, corresponding to the number 19.*

13. 13 + 13 = *26*   -1 + 5 = *4*   -1 + 26 = *25*

13a. Added marks: *On both the C and D scales, add 1 unit of distance to left of 0 and call it -1; add 1 unit of distance to the right of 25 and call it 26.*

13b. Invented problem exceeding capacity of slide rule: 14 + 14 = ____. *(Any problem with an answer less than -1 or greater than 26 is appropriate.)*

NOTES (next page)

16. Notice that these subtraction problems always *start* by moving the hairline, and *continue* by moving the C-scale.

17. Summarize addition and subtraction on your board with this rule:

Addition: start with index; continue with hairline.
Subtraction: start with hairline; continue with C-scale.

Point out that using a slide rule to compute a long string of numbers is a little like dancing! This involves *alternating* hairline *addition* and C-scale *subtraction*. If the plus and minus signs don't alternate, simply add or subtract zeros as necessary. This keeps the dance alternating, and doesn't change a thing.

☞ *See GLASToids supplement, pg 17.*

18. Every lesson ends in a "GLASToid." These questions tie NASA's GLAST mission into lesson concepts as much as possible, or present interesting factoids along the way. All necessary information to answer every GLASToid is hidden somewhere within this supplement. These puzzles are generally fun and easy do.

ANSWERS (next page)

15. First... *Set the Hairline over "13" on the D scale.*
Second... *Slide the C scale to "5" under the Hairline.*
Third... *Read "8," under the Index on the D scale.*

16.
10 - 4 = *6*	9 - 4 = *5*	17 - 0 = *17*	2.5 - 1.5 = *1.0*
18 - 11 = *7*	10 - 6 = *4*	15 - 15 = *0*	2.2 - 1.8 = *0.4*
25 - 24 = *1*	12 - 10 = *2*	5 - 3 = *2*	4.0 - 1.7 = *2.3*

17.
8 + 5 - 3 = *10*	23 - 21 = *2*
5 + 16 - 20 = *1*	20 - 5 + 6 = *21*
3 + 19 - 12 = *10*	12 - 24 + 16 - 2 = *2*
4 + 9 - 6 + 12 = *19*	9 - 8 + 7 - 6 + 3 = *5*
20 + 5 - 10 + 3 - 6 = *12*	8 - 2 + 0 - 2 + 0 - 2 = *2*
1 + 2 - 0 + 3 - 0 + 4 = *10*	20 - 5 - 5 - 5 = *5*
1 + 3 + 5 + 7 + 9 = *25*	24 - 6 - 6 - 6 - 6 = *0*

When solving combination problems, it is helpful to add or subtract zeros *so the +'s and -'s continue to alternate.* (Adding 0 with the hairline frees the C-scale to make the next subtraction. Subtracting 0 with the C-scale frees the hairline to make the next addition.)

GLASToids!

18a. NASA stands for the *National Aeronautics and Space Administration.*

18b. GLAST stand for the *Gamma-ray Large Area Space Telescope.*

18c. Students should use their slide rules to *subtract their current calendar year from the 2006 launch date.*

NOTES: **Activity A1**

ADDING SLIDE RULE

Activity A1

14. Try this **subtraction** problem: $\overset{H}{\underset{D}{13}} - \overset{C}{\underset{H}{5}} = \overset{I}{\underset{D}{8}}$

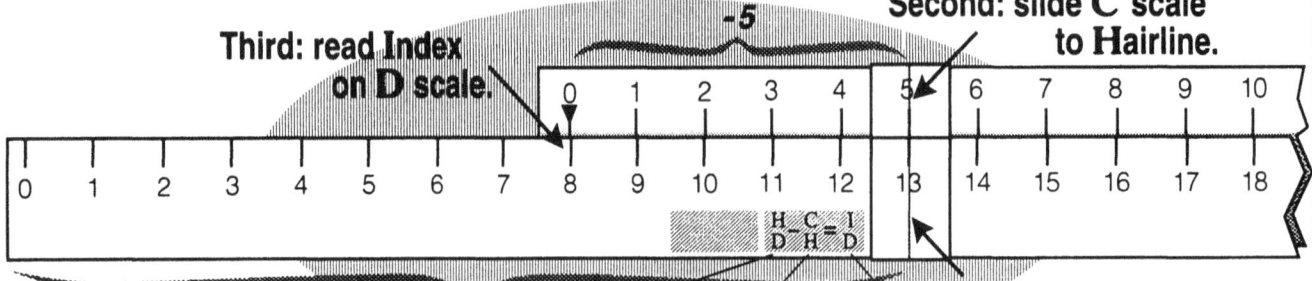

15. Find the grey "–" box on your slide rule. Summarize how the letters help you **subtract**: 13 – 5 = 8

First ...	Second...	Third...

16. Find these **differences** on your slide rule.

$\overset{H}{\underset{D}{10}} - \overset{C}{\underset{H}{4}} = $ ____ $\underset{D}{I}$	$\overset{H}{\underset{D}{9}} - \overset{C}{\underset{H}{4}} = $ ____ $\underset{D}{I}$	$\overset{H}{\underset{D}{17}} - \overset{C}{\underset{H}{0}} = $ ____ $\underset{D}{I}$	$\overset{H}{\underset{D}{2.5}} - \overset{C}{\underset{H}{1.5}} = $ ____ $\underset{D}{I}$
18 – 11 = ____	10 – 6 = ____	15 – 15 = ____	2.2 – 1.8 = ____
25 – 24 = ____	12 – 10 = ____	5 – 3 = ____	4.0 – 1.7 = ____

17. Try these combination problems:

$\overset{I}{\underset{D}{8}} + \overset{H}{\underset{C}{5}} - \overset{C}{\underset{H}{3}} = $ ____ $\underset{D}{I}$ $\overset{H}{\underset{D}{23}} - \overset{C}{\underset{H}{21}} = $ ____ $\underset{D}{I}$

5 + 16 – 20 = ____   20 – 5 + 6 = ____

3 + 19 – 12 = ____   12 – 24 + 16 – 2 = ____

4 + 9 – 6 + 12 = ____   9 – 8 + 7 – 6 + 3 = ____

20 + 5 – 10 + 3 – 6 = ____   8 – 2 + 0 – 2 + 0 – 2 = ____

1 + 2 – 0 + 3 – 0 + 4 = ____   20 – 5$\overset{+0}{\wedge}$– 5$\overset{+0}{\wedge}$– 5 = ____

1 + 3$\overset{-0}{\wedge}$+ 5$\overset{-0}{\wedge}$+ 7$\overset{-0}{\wedge}$+ 9 = ____   24 – 6 – 6 – 6 – 6 = ____

Why is it sometimes helpful to add or subtract 0 when computing a string of numbers on a slide rule?

18. Get the *GLASToids* page. Find the answers to these questions:

   a. These lessons are brought to you by NASA. What does the acronym NASA stand for? ____

   b. NASA plans to launch GLAST into earth orbit in the fall of 2006. What does GLAST stand for? ____

   c. Use your slide rule to calculate the number of years before (or after) launch date.

Make six PLASTIC HAIRLINES   Supplement to Activity A1

1. Cut out the large grey rectangle. Fold it along the center dotted line.

2. Paper clip *two* layers of plastic *inside* the folded paper.

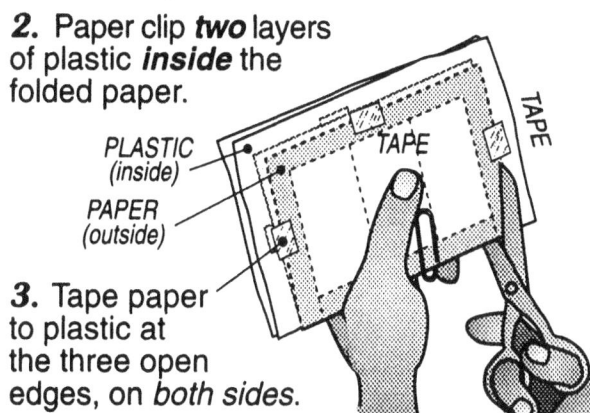

PLASTIC (inside)
PAPER (outside)
TAPE

3. Tape paper to plastic at the three open edges, on *both sides*.

4. Holding this "sandwich" together firmly, cut away the grey borders along the three open sides.

5. Secure the cut edges with three paper clips. Remove the original clip, then cut away the folded grey side.

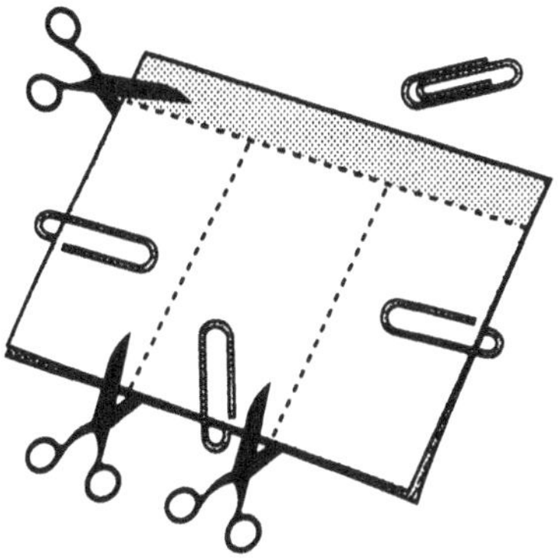

6. Cut the remaining white rectangle into three equal pieces along the dashed lines.

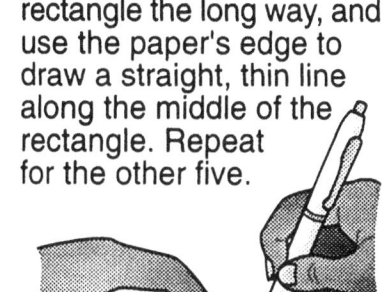

7. Remove the paper clips, and separate the plastic. (Rub a corner between your fingers if layers stick.)

8. Find a ballpoint pen that writes well on a plastic scrap. (Don't scribble on your good rectangles.)

9. Mark your *hairlines*: fold one of the *paper* rectangles in half the long way. Insert a *plastic* rectangle the long way, and use the paper's edge to draw a straight, thin line along the middle of the rectangle. Repeat for the other five.

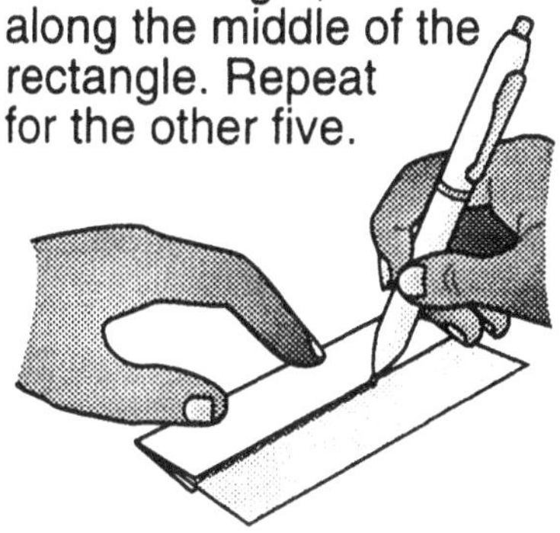

10. Fold a piece of scratch paper (hide printing inside), in quarters to make a pocket. Write your name on the outside.

11. Paper clip your six hairlines inside this pocket until needed.

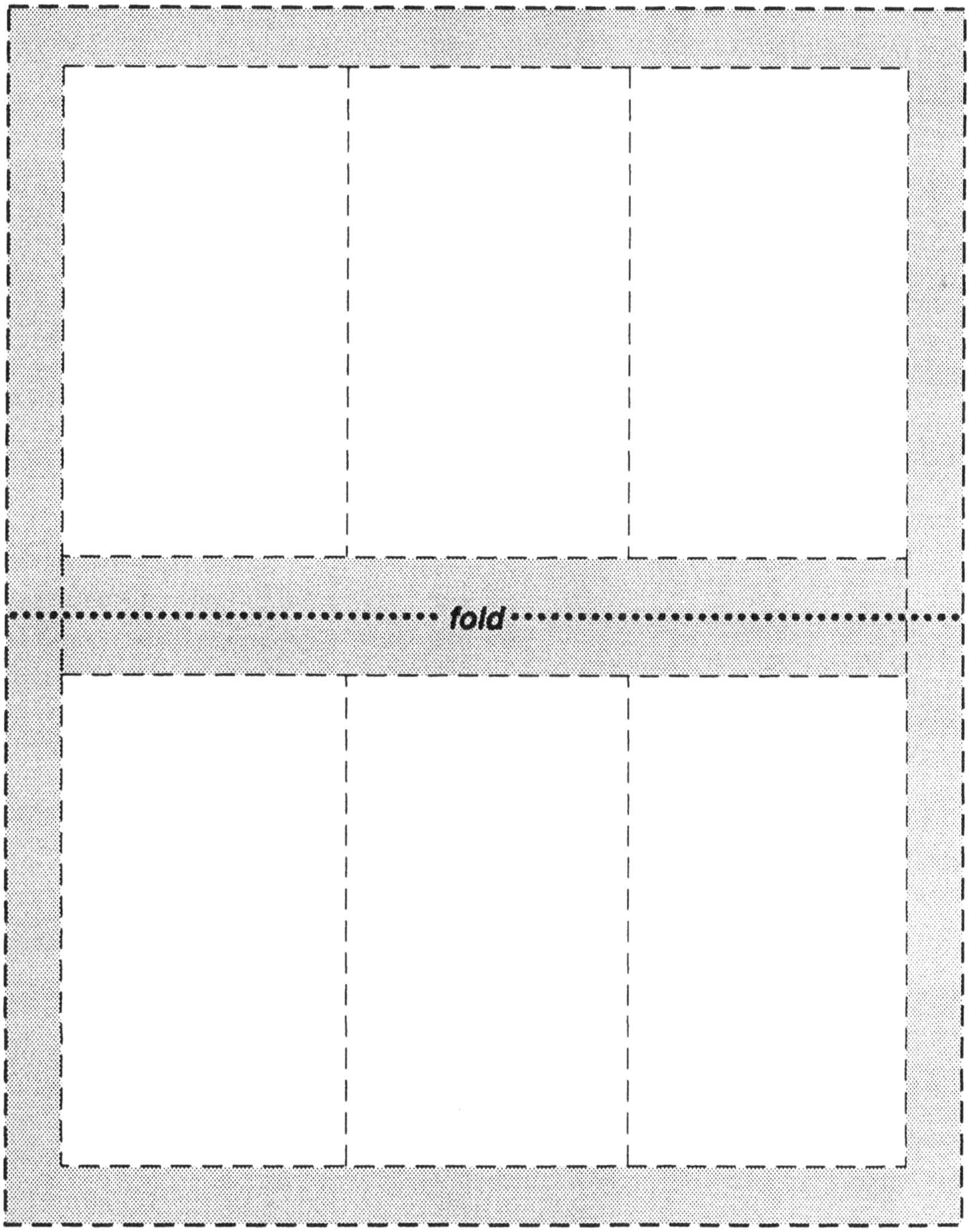

Copyright © 2003 by TOPS Learning Systems, Canby OR 97013.

PAGE 15

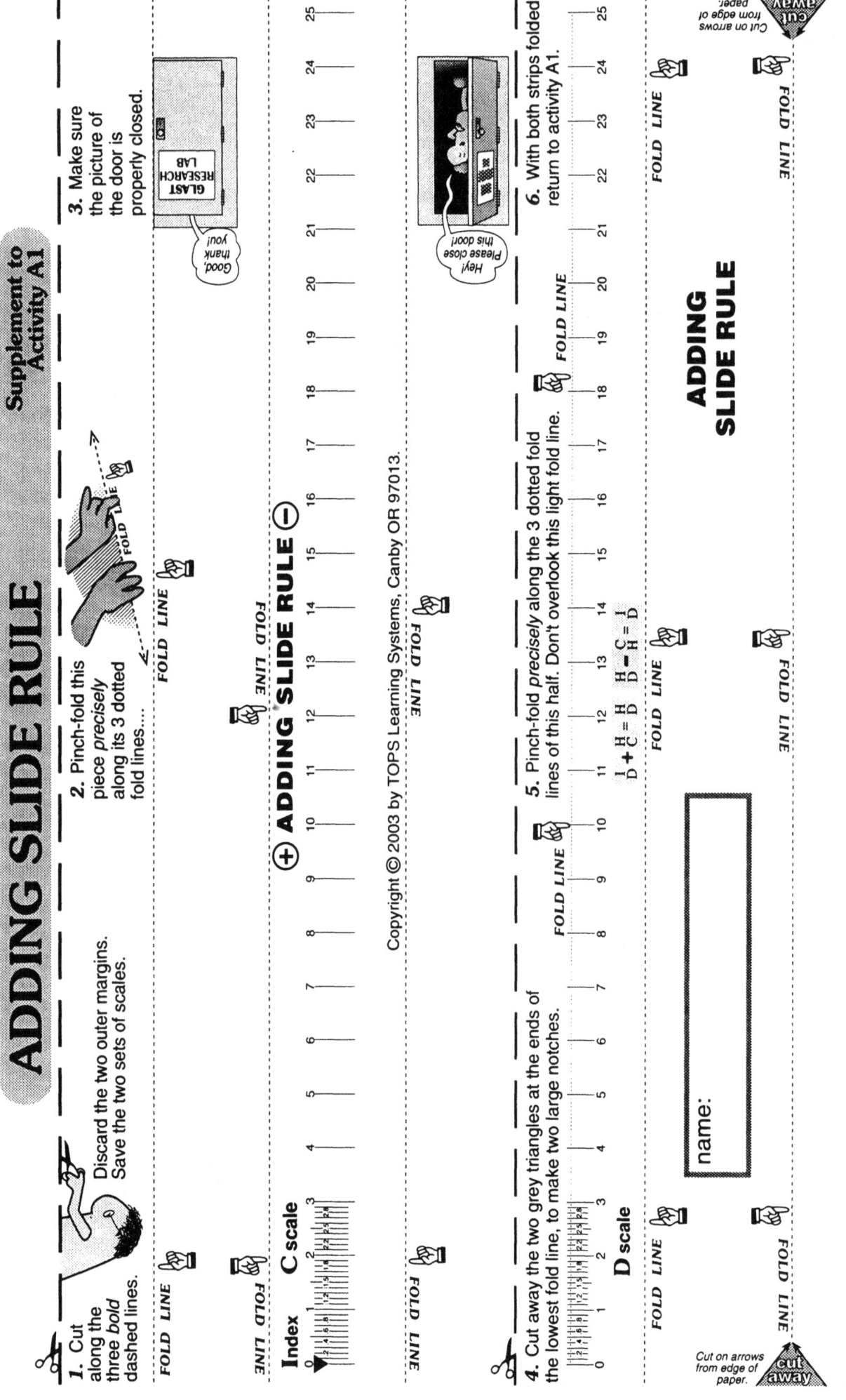

MULTIPLYING SLIDE RULE

Activity A2
1 of 3 pages

1. Get the *Multiplying Slide Rule* supplement, and follow directions as before.

2. Fit the C-scale fully inside the D-scale. Wrap with a hairline, taping so both ends of the line perfectly match.

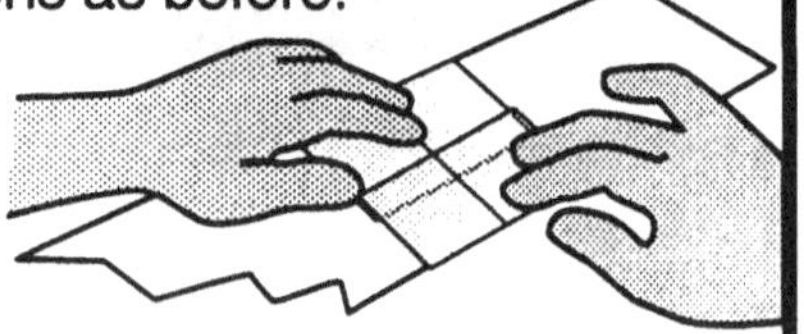

3. Try these multiplication problems on your slide rule.

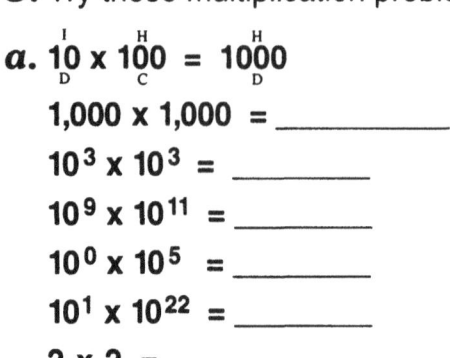

a. $\underset{D}{\overset{I}{10}} \times \underset{C}{\overset{H}{100}} = \underset{D}{\overset{H}{1000}}$

1,000 × 1,000 = _____

$10^3 \times 10^3$ = _____

$10^9 \times 10^{11}$ = _____

$10^0 \times 10^5$ = _____

$10^1 \times 10^{22}$ = _____

2 × 2 = _____

3 × 2 = _____

30 × 2 = _____

b. Describe operationally (say what you do), to multiply numbers on this slide rule. Use slide rule vocabulary.

c. In summary, this slide rule multiplies numbers by...

4. Try these division problems on your slide rule.

a. $\underset{D}{\overset{H}{1,000,000}} \div \underset{H}{\overset{C}{1,000}} = \underset{D}{\overset{I}{1,000}}$

10,000 ÷ 100 = _____

$10^4 \div 10^2$ = _____

$10^{10} \div 10^3$ = _____

$10^{20} \div 10^{19}$ = _____

$10^7 \div 10^0$ = _____

10 ÷ 2 = _____

600 ÷ 30 = _____

30 ÷ 3 = _____

b. Describe operationally how to divide numbers on this slide rule. Use slide rule vocabulary.

c. In summary, this slide rule divides numbers by...

5. Evaluate on your slide rule:

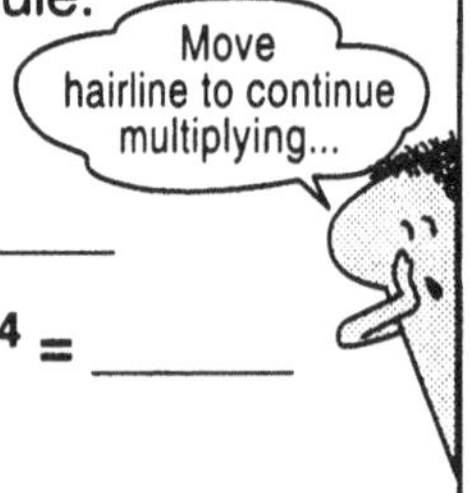

$10^2 \times 10^2$ = _____

$10^3 \times 10^3 \overset{+1}{\wedge} \times 10^3$ = _____

$10^4 \times 10^4 \overset{+1}{\wedge} \times 10^4 \overset{+1}{\wedge} \times 10^4$ = _____

$(10^5)^5$ = _____

Move hairline to continue multiplying...

When raising "ten to a power" to yet another power, what operation do you do with the exponents?

6. Evaluate on your slide rule:

$10^{19} \div 10^{17}$ = _____

$10^{15} \div 10^5 \overset{\times 1}{\wedge} \div 10^5 \overset{\times 1}{\wedge} \div 10^5$ = _____

$10^{24} \div 10^8 \overset{\times 1}{\wedge} \div 10^8 \overset{\times 1}{\wedge} \div 10^8$ = _____

$\sqrt[3]{10^{24}}$ = _____

Move C scale to continue dividing...

When taking the root of "ten to a power," what operation do you do with the index over the radical and the exponent under it?

continue next page

OVERVIEW / OBJECTIVES

Students will construct *Multiplying Slide Rules* scaled in base-10 exponents and use them to calculate products and quotients. They will come to appreciate that "super numbers" (exponents, orders of magnitude, and logarithms) play by different rules of arithmetic than "ordinary numbers" (numbers, powers of ten, and antilogs).

TIME: 1 hour and 25 minutes.

INTRODUCTION

Can anyone imagine how adding and subtracting ruler distances might multiply and divide numbers? These rulers couldn't be scaled in ordinary numbers, because we've seen that such numbers add and subtract. But what about exponents? We know from our study of algebra that exponents **add to multiply**, and **subtract to divide!**

Fix a meter stick on a wall with a movable meter stick above it, as before. Observe that each meter is divided into 10 decimeters. Let each decimeter represent a power of ten: 0 cm = 10^0, 10 cm = 10^1, 20 cm = 10^2, ..., 100 cm = 10^{10}. (Apply appropriate tape labels.) Now count with your class up the decimeter scale, beginning with 10^0 = 1: one, ten, hundred, thousand, ten thousand, hundred thousand, million, ten million, hundred million, billion, ten billion.

Slide the the left end of the movable meter (10^0) to 10^2 (the 2 decimeter position). Using your hand as a hairline, show how numbers top and bottom, representing exponents, now add to multiply and subtract to divide.

multiply: $10^2 \times 10^3 = 10^5$
100 x 1,000 = 100,000
divide: $10^5 \div 10^3 = 10^2$
100,000 ÷ 1,000 = 100

Scales calibrated in exponents (instead of ordinary numbers), are said to be logarithmic. Why? Because *logs are exponents!* (Write this on your board.)

NOTES (this page)

1-2. Students should construct the *Multiplying Slide Rule* as they did the *Adding Slide Rule*, wrapping it with one of their *Plastic Hairlines*. When most have completed this task, ask students to pause and compare their new and old slide rules. List important differences on your board:

How is the Multiplying Slide Rule different?

a. It's calibrated in base 10 exponents (not ordinary numbers).

b. Each major division on the slide rule now jumps by an order of magnitude (1, 10, 100, 1000...) instead of by the next whole number (0, 1, 2, 3...).

c. Subdivisions are bunched, in a regular repeating pattern.

d. The grey boxes show parallel operations, except what previously added now multiplies; what previously subtracted now divides.

3. Multiplication always begins by setting the index, and continues by moving the hairline.

4. Division always begins by setting the hairline, and continues by moving the C scale.

ANSWERS (this page)

3a. 10 x 100 = 1000
1,000 x 1,000 = *1,000,000*
$10^3 \times 10^3 = 10^6$
$10^9 \times 10^{11} = 10^{20}$
$10^0 \times 10^5 = 10^5$
$10^1 \times 10^{22} = 10^{23}$
2 x 2 = *4*
3 x 2 = *6*
30 x 2 = *60*

3b. To multiply, *move the index over the first number on the D scale, then slide the hairline over the second number on the C scale. Read the product under the hairline on the D scale.*

3c. This slide rule multiplies numbers by *adding exponents*.

4a. 1,000,000 ÷ 1,000 = 1,000
10,000 ÷ 100 = *100*
$10^4 \div 10^2 = 10^2$
$10^{10} \div 10^3 = 10^7$
$10^{20} \div 10^{19} = 10^1$
$10^7 \div 10^0 = 10^7$
10 ÷ 2 = *5*
600 ÷ 30 = *20*
30 ÷ 3 = *10*

4b. To divide, *move the hairline over the first number on the D scale, then slide the the second number on the C scale under the hairline. Read the quotient under the index on the D scale.*

4c. This slide rule divides numbers by *subtracting exponents*.

5. $10^2 \times 10^2 = 10^4$
$10^3 \times 10^3 \times 10^3 = 10^9$
$10^4 \times 10^4 \times 10^4 \times 10^4 = 10^{16}$
$(10^5)^5 = 10^{25}$

When raising "ten to a power" to yet another power, *the exponents multiply* (add successively on the slide rule).

6. $10^{19} \div 10^{17} = 10^2$
$10^{15} \div 10^5 \div 10^5 \div 10^5 = 10^0$
$10^{24} \div 10^8 \div 10^8 \div 10^8 = 10^0$
$\sqrt[3]{10^{24}} = 10^8$

When taking the root of "ten to a power," *the index divides into the exponent* (the exponent of the root subtracts successively to zero, yielding the same number of equal parts as the index).

MULTIPLYING SLIDE RULE

7. Using a sharp pencil, lightly **circle** the base-ten **exponents** from 0 to 6 on both scales of your slide rule.

8. These circled exponents are called **orders of magnitude** (OM's).

   a. Write the number value, in words, of each OM you circled:

   ___one___, ___ten___, _____, _____,

   _____, _____, _____,

   b. What number is...

     3 OM's larger than a million? _____

     9 OM's larger than a thousand? _____

     12 OM's larger than a trillion? _____

Americans say "billion." The English say "milliard."
Americans say "trillion." The English say "billion."

   c. This *Multiplying Slide Rule* spans _____ orders of magnitude, while your *Adding Slide Rule* spans less than _____ orders of magnitude. Account for this difference. →

9. The exponents you circled are also called **logarithms**. Follow the pattern to complete this

LOG Table

log 10^exp	=	(OM) (exp)	=	log N
log 10^6	=	⑥	=	log 1,000,000
log 10^5	=	⑤	=	log 100,000
	=	④	=	
	=	○	=	
	=	○	=	
	=	○	=	
	=	○	=	
	=	○	=	
	=	○	=	
	=	○	=	
	=	○	=	

10. How do base-10 exponents (logs), determine decimal place? Invent a rule:

$$\text{log } 1 = 0$$
$$\downarrow$$
$$0001\overset{.}{}000000$$

11. Exponents, orders of magnitude, and logs are all the same.

1. **EXPONENTS** (exp)
2. **ORDERS OF MAGNITUDE** (OM)
3. **LOGARITHMS** (log N)

I have 3 names! *They're all equal!*

$10^{⑨}$ ← What is this? Describe it fully!

NOTES (this page)

9. Only base-10 exponents (common logs) function as orders of magnitude. This is not true for logs in other bases.

The *Multiplying Slide Rule* starts with $10^0 = 1$. But as the table suggests, its logarithmic scale could be extended leftward: *exp = 0, exp = -1, exp = -2,* ... Review with your class how negative exponents (logs) translate into fractions:

$$10^{-1} = 1/10^1 = 1/10 = 0.1$$

This log table is saying something very important. Write this generalization on your board and ask students to memorize it:

The (log of a number) is an (exponent.)
log N = exp

(See notes, step 13, for the inverse statement.)

ANSWERS (this page)

7. Students should *circle* base-10 exponents on both scales, from $10^{⓪}$ out to $10^{⑥}$ (14 circles altogether).

8a. *one, ten, hundred, thousand, ten thousand, hundred thousand, million.*

8b. *A billion, or 10^9, is 3 OM's larger than a million.*
A trillion, or 10^{12}, is 9 OM's larger than a thousand.
A trillion trillion, or 10^{24}, is 12 OM's larger than a trillion.

8c. *This Multiplying Slide Rule spans 25 orders of magnitude, while your Adding Slide Rule spans less than 2 orders of magnitude.*

This difference results from different calibrations. The adding scales are linear, increasing 1 number at a time. The multiplying scales are logarithmic (calibrated in exponents), increasing 1 order of magnitude at a time.

9. $\log 10^6 = ⑥ = \log 1{,}000{,}000$
$\log 10^5 = ⑤ = \log 100{,}000$
$\log 10^4 = ④ = \log 10{,}000$
$\log 10^3 = ③ = \log 1{,}000$
$\log 10^2 = ② = \log 100$
$\log 10^1 = ① = \log 10$
$\log 10^0 = ⓪ = \log 1$
$\log 10^{-1} = ⊖1 = \log .1$
$\log 10^{-2} = ⊖2 = \log .01$
$\log 10^{-3} = ⊖3 = \log .001$
$\log 10^{-4} = ⊖4 = \log .0001$

10. *The value of the base-10 exponent (log) tells you how many places to move the decimal from the standard "ones" position. If the log is positive, move it right that many places; if negative, move it left.*

11. *This "9" is a base-10 exponent, or log. It also functions as an order of magnitude.*

NOTES (next page)

13. The "antilog exp" notation looks strange when seen for the first time, but it's just another way of saying "10^{exp}."

This antilog table is saying something very important. Write this generalization on your board and ask students to memorize it:

The [antilog of an exponent] is a [number.]
antilog exp = N

ANSWERS (next page)

12. Students should *box* powers of ten (and their decimal equivalents), on the D scale only, (14 boxes altogether).

13. $\boxed{1{,}000{,}000} = 10^6 = $ antilog 6
$\boxed{100{,}000} = 10^5 = $ antilog 5
$\boxed{10{,}000} = 10^4 = $ antilog 4
$\boxed{1{,}000} = 10^3 = $ antilog 3
$\boxed{100} = 10^2 = $ antilog 2
$\boxed{10} = 10^1 = $ antilog 1
$\boxed{1} = 10^0 = $ antilog 0
$\boxed{.1} = 10^{-1} = $ antilog -1

14. *"10^9" is a power of ten that functions like an ordinary number (1 billion). It is also called an antilog.*

15. ②+②=④ $100 \times 100 = 10{,}000$
①+④=⑤ $10 \times 10{,}000 = 100{,}000$
③+①=④ $1{,}000 \times 10 = 10{,}000$
⑤-③=② $100{,}000 / 1{,}000 = 100$
⑥-②=④ $1{,}000{,}000 / 100 = 10{,}000$
③-⓪=③ $1{,}000 / 1 = 1{,}000$

When super numbers *add*, ordinary numbers *multiply*.
When super numbers *subtract*, ordinary numbers *divide*.

GLASToids!

16a. GRB stands for *Gamma Ray Burst.*
16b. Glast is expected to have a lifetime of *5 years.*
16c. *200 GRBs/year × 5 years = 1,000 GRBs.*

NOTES: **Activity A2**

MULTIPLYING SLIDE RULE

Activity A2

12. Using a sharp pencil, lightly BOX numbers on the D scale. Do this for **powers of ten** through 10^6, and **numbers** through 1,000,000.

13. These boxed numbers are also called *antilogs*. Follow the pattern to complete this

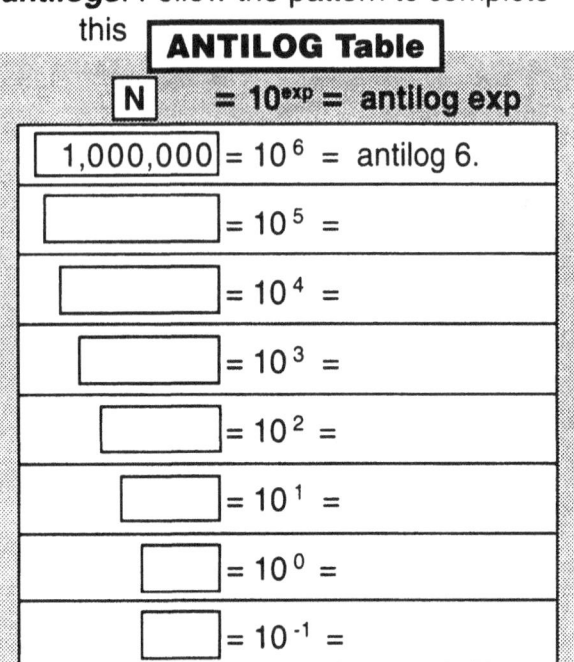

ANTILOG Table

N	= 10^{exp}	= antilog exp
1,000,000	= 10^6	= antilog 6.
	= 10^5	=
	= 10^4	=
	= 10^3	=
	= 10^2	=
	= 10^1	=
	= 10^0	=
	= 10^{-1}	=

14. Numbers, powers of ten, and antilogs all have the same ordinary properties.

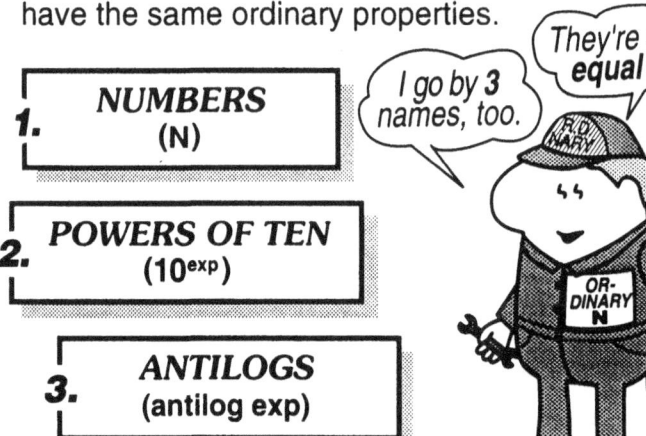

1. **NUMBERS** (N)
2. **POWERS OF TEN** (10^{exp})
3. **ANTILOGS** (antilog exp)

I go by 3 names, too. *They're all equal!*

10^9 What is this? Describe it fully!

15. Compare "super" and "ordinary" number operations on your slide rule.

SUPER (logs = exps)	ORDINARY (numbers)
② + ② = ④	100 x 100 = **10,000**
① + ④ =	
	1000 x 10 =

SUPER (logs = exps)	ORDINARY (numbers)
⑤ − ③ = ②	100,000 / 1,000 =
	1,000,000 / 100 =
③ − ⓪ =	

When super numbers ___add___,

ordinary numbers _____.

When super numbers _____,

ordinary numbers _____.

GLASToids

16. Ponder the *GLASToids* page. You'll need *one* for each question.

a. What's a GRB? _____

b. How long is the GLAST mission expected to last? _____

c. If 200 GRBs are "seen" by GLAST per year, how many events might be documented over the life of the mission? (Work this problem on your slide rule!)

MULTIPLYING SLIDE RULE
Supplement to Activity A2

1. Cut along the three *bold* dashed lines.

Discard the two outer margins. Save the two sets of scales.

FOLD LINE

2. Pinch-fold this piece *precisely* along its 3 dotted fold lines....

FOLD LINE

3. Make sure the picture of the door is properly closed.

Good, thank you!

GLAST RESEARCH LAB

10^{25} 10^{24} 10^{23} 10^{22} 10^{21} 10^{20} 10^{19} 10^{18} 10^{17} 10^{16} 10^{15} 10^{14} 10^{13} 10^{12} 10^{11} 10^{10} 10^9 10^8 10^7 10^6 1,000,000 10^5 100,000 10^4 10,000 10^3 1,000 10^2 100 10^1 10 10^0 1

Index C scale

FOLD LINE

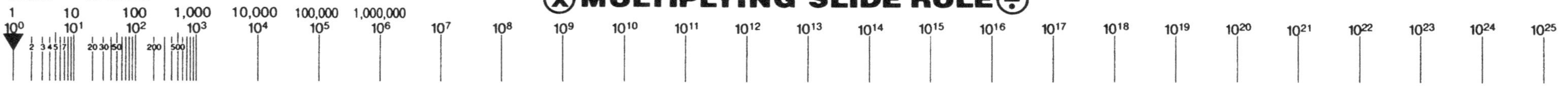

(×) **MULTIPLYING SLIDE RULE** (÷)

FOLD LINE

Copyright © 2003 by TOPS Learning Systems, Canby OR 97013.

FOLD LINE

10^{25} 10^{24} 10^{23} 10^{22} 10^{21} 10^{20} 10^{19} 10^{18} 10^{17} 10^{16} 10^{15} 10^{14} 10^{13} 10^{12} trillion 10^{11} 10^{10} 10^9 billion 10^8 10^7 10^6 million 1,000,000 10^5 100,000 10^4 10,000 10^3 1,000 thousand 10^2 100 10^1 10 10^0 1

D scale

Hey! Please close this door!

6. With both strips folded, return to activity A2.

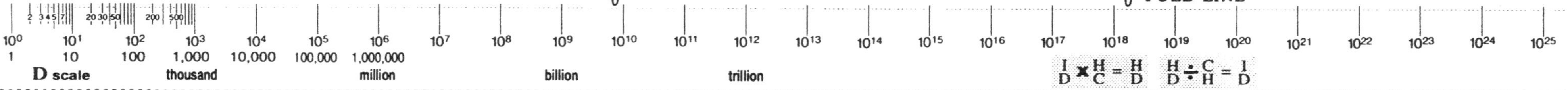

5. Pinch-fold *precisely* along the 3 dotted fold lines of this half. Don't overlook this light fold line.

$I \times H = H \quad H \div C = I$
$D \times C = D \quad D \div H = D$

MULTIPLYING SLIDE RULE

name:

4. Cut away the two grey triangles at the ends of the lowest fold line, to make two large notches.

FOLD LINE

Cut on arrows from edge of paper. **cut away**

Cut on arrows from edge of paper. **cut away**

LOG TAPE

Activity B1

1. Get the *Base-Ten Log Tape* supplement. Cut precisely into 4 equal strips (especially along the ends). Tape these strips end to end, carefully matching numbers from low to high.

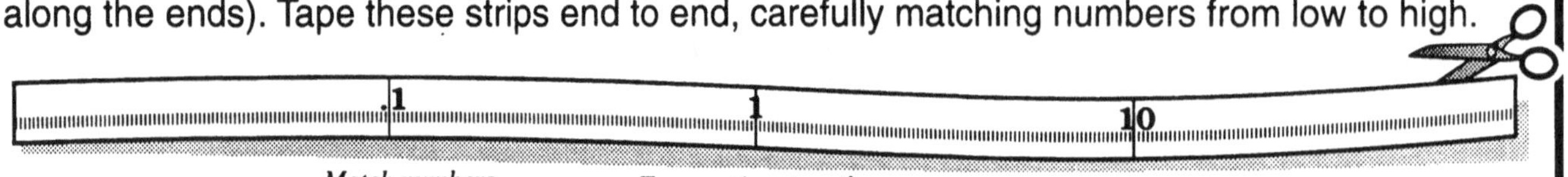

Match numbers carefully — *Tape sections together*

Those little **bars** are another way of writing negative exponents:
$\overline{2}.301$
Under the **bar** is **negative**... ...the decimal part is **positive**.

2a. Consult your tape to complete each **number equation**.

| $.02 = 10^{\overline{2}.301}$ | $.2 =$ | $2 =$ | $20 =$ |

2b. Take the log of both sides. Notice how this "pulls down" the exponent, changing each number equation into a **log equation**.

| $\log .02 = \log 10^{\overline{2}.301}$ $\log .02 = \overline{2}.301$ | $\log .2 = 10^{\overline{1}.301}$ | | |

3. Practice writing antilogs **3 equal ways**, and logs **2 equal ways**. Follow the patterns!

4. Study your table to answer these questions:

ANTILOGS			LOGS	
N (Number)	10^{exp} (Power of Ten)	antilog of exp	log of N	exp
3	$10^{0.477}$	antilog 0.477	log 3	0.477
		antilog 0.602		0.602
6.5				0.813
38				
100				
			log 10	
		antilog 0		
				-1
				$\overline{2}.000$
			log .045	$\overline{2}.653$
			log .0045	
	$10^{1.903}$			
	$10^{2.903}$			
95				
950				
.095				
				0.114
				2.114
				$\overline{2}.114$

a. Antilogs are expressed in three different ways, as...

_____ ,

and... _____ ,

and... _____ .

b. Logs are expressed in two different ways...

_____ ,

and... _____ .

c. The antilog of an exponent equals...

_____

and... _____ .

d. The log of a number equals...

_____ .

e. The antilog of the log of a number is...

_____ .

f. The log of the antilog of an exponent is...

_____ .

continue

OVERVIEW / OBJECTIVES

Students will construct *Log Tapes* calibrated in base-ten exponents, then use them to derive relationships between base-ten logs (exponents) and antilogs (ordinary numbers).

TIME: 1 hour and 30 minutes.

INTRODUCTION

★ Make the *Log Tape*, as directed. Stick it to your board with small rolls of masking tape. Then label some of its logs and antilogs (see the illustration below) within the context of this 6-part class discussion:

a: The bottom scale on this *Log Tape* is evenly calibrated in *base-ten exponents*. I'll label a few of these underneath, so you can see them from your desk:

$$-2, \quad -1, \quad 0, \quad 1, \quad 2$$

b: Now let's write the *number* values for each of these exponents directly above, both as powers of ten and as decimals:

$$.01, \quad .1, \quad 1, \quad 10, \quad 100$$
$$10^{-2}, \quad 10^{-1}, \quad 10^0, \quad 10^1, \quad 10^2$$

Notice how these base-ten exponents increase linearly, by adding 1, while the corresponding numbers increase logarithmically, by multiplying by 10.

c: Since *logs are exponents* (log N = exp), we can write this alternate notation underneath the exponents."

$$\log .01, \quad \log .1, \quad \log 1, \quad \log 10, \quad \log 100$$

d: And since *antilogs are numbers* (antilog exp = N), we can write this alternate notation above the numbers.

antilog -2, antilog -1, antilog 0, antilog 1, antilog 2

e: Exponents don't have to be *whole* numbers! Let's calibrate halfway between the exponent 0 and 1, and label it five ways, as before:

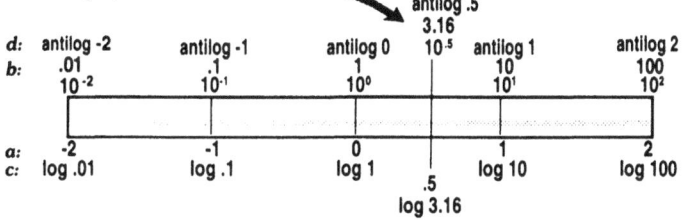

f: Label the 3 remaining half-way positions below the log scale: (-2, **-1.5**, -1, **-.5**, 0, .5, 1, **1.5**, 2) and above the number scale:

(.01, **.0316**, .1, **.316**, 1, 3.16, 10, **31.6**, 100).

Has anyone seen a scale where *3* is located almost half-way to *10*, and *30* is almost halfway to *100*? (Examine subcalibrations on the *Multiplying Slide Rule*. This repeating "bunching" pattern is typical of base-10 logarithmic scales.)

★ Negative exponents can be written two ways, both of which are shown on the *Log Tape* to the left of 0 (log 1). One way simply expresses the complete log as a negative quantity. Another form uses a **bar** over the *whole number part* (characteristic) so the *decimal part* (mantissa) always remains positive. In this form it is easy to see that the mantissa of .02, .2, 2, and 20 is always .301! Only the characteristic, which determines the decimal place, changes. Thus, while 2 has a characteristic of 0, .2 has a characteristic of $\bar{1}$, and .02 has a characteristic of $\bar{2}$.

NOTES (this page)

2. Students should copy each base-10 exponent as printed on the *Log Tape*. Notice that these same exponents can also be read from the finely calibrated log scale underneath.

3. This table helps students understand that there are 3 equal ways to write antilogs (numbers) and 2 equal ways to write logs (exponents). Filling it in is really quite easy. If necessary, complete the first few lines with your whole class together, to overcome fears of this new terminology.

ANSWERS (this page)

2a. $.02 = 10^{\bar{2}.301}$ $.2 = 10^{\bar{1}.301}$ $2 = 10^{0.301}$ $20 = 10^{1.301}$

2b. $\log .02 = \log 10^{\bar{2}.301}$ $\log 2 = \log 10^{0.301}$
$\log .02 = \bar{2}.301$ **log 2 = 0.301**

$\log .2 = \log 10^{\bar{1}.301}$ $\log 20 = \log 10^{1.301}$
log .2 = $\bar{1}$.301   **log 20 = 1.301**

3.
3	=	$10^{0.477}$	=	antilog 0.477	log 3	= 0.477
4	=	$10^{0.602}$	=	antilog 0.602	**log 4**	= 0.602
6.5	=	$10^{0.813}$	=	antilog 0.813	**log 6.5**	= 0.813
38	=	$10^{1.580}$	=	antilog 1.580	**log 38**	= 1.580
100	=	10^2	=	antilog 2	log 100	= 2
10	=	10^1	=	antilog 1	log 10	= 1
1	=	10^0	=	antilog 0	**log 1**	= 0
.1	=	10^{-1}	=	antilog -1	**log .1**	= -1
.01	=	10^{-2}	=	antilog -2	**log .01**	= -2.000
.045	=	$10^{\bar{2}.653}$	=	antilog $\bar{2}.653$	log .045	= $\bar{2}.653$
.0045	=	$10^{\bar{3}.653}$	=	antilog $\bar{3}.653$	log .0045	= $\bar{3}.653$
80	=	$10^{1.903}$	=	antilog 1.903	log 80	= 1.903
800	=	$10^{2.903}$	=	antilog 2.903	**log 800**	= 2.903
95	=	$10^{1.978}$	=	antilog 1.978	**log 95**	= 1.978
950	=	$10^{2.978}$	=	antilog 2.978	**log 950**	= 2.978
.095	=	$10^{\bar{2}.978}$	=	antilog $\bar{2}.978$	**log .095**	= $\bar{2}.978$
1.3	=	$10^{0.114}$	=	antilog 0.114	**log 1.3**	= 0.114
130	=	$10^{2.114}$	=	antilog 2.114	**log 130**	= 2.114
.013	=	$10^{\bar{2}.114}$	=	antilog $\bar{2}.114$	**log .013**	= $\bar{2}.114$

4a. Antilogs: **numbers, powers of ten, antilogs.**

4b. Logs: **logs, exponents.**

4c. The antilog of an exponent = **a number, a power of ten.**

4d. The log of a number = **an exponent.**

4e. The antilog of the log of a number is a **number**.
(Substitute log N for exp in 3rd column.)

4f. The log of the antilog of an exponent is an **exponent**.
(Substitute antilog exp for N in 4th column.)

LOG TAPE

Activity B1

5. Imagine adding 2 more sections onto each end of your *Log Tape*, then compressing it and turning it sideways, as illustrated.

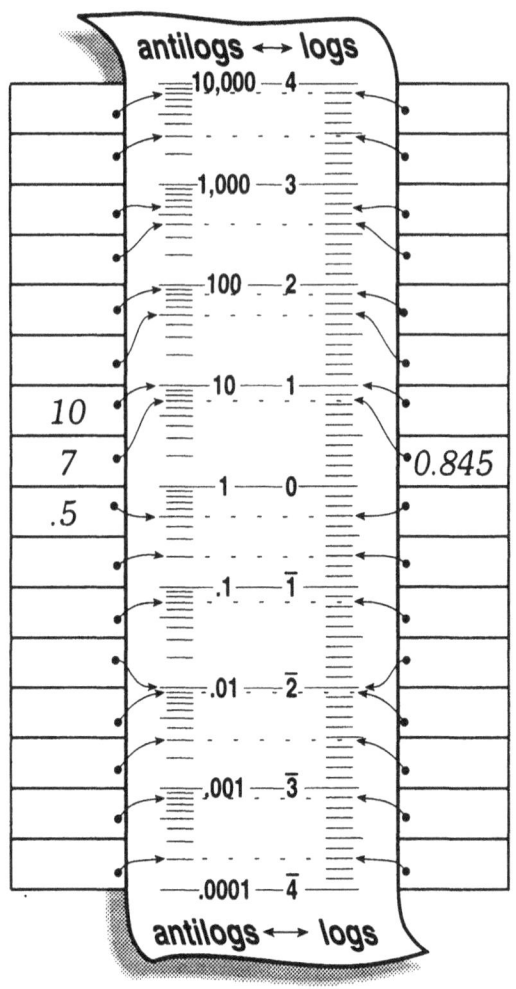

a. Write in each **antilog** on the left side (at the tiny arrows), beginning with "7." This corresponds to "7" on your log tape.

b. On the right side, write the **log** of each number you wrote on the left. Read to 2 decimal places on this compressed scale, then consult your more detailed *Log Tape* for accuracy to 3 decimal places.

c. This compressed scale contains 8 repeating cycles (count them). How many repeating cycles does your *Log Tape* contain?
_____

d. As the log of a number gets "very" negative, what happens to the number itself?
_____

e. If a number is greater than zero but less than one, what can you say about the log of this number?
_____

f. A "bar-log" is an alternate way to write a negative log. The digit printed under the bar is *negative*, but the decimal numbers that follow it are *positive*. To write an "unbarred" negative log, take the difference between the positive and negative parts. (Or, read "unbarred" logs directly from your *Log Tape*, printed below the bar logs.)

$\bar{1}.699 = -1.000 + .699 = .301$ $\bar{2}.778 =$

$\bar{1}.301 =$ $\bar{2}.477 =$

6. For each **antilog** range, give the corresponding **log** range.

   1 to 10: _____
   10 to 100: _____
   1 to .1: _____
   .1 to .01: _____

7. For each **log** range, give the corresponding **antilog** range.

   −3 to −4: _____
   −4 to −5: _____
   3 to 4: _____
   4 to 5: _____

8. Over what energy range will GLAST detect gamma rays? Search the glastoids!

Express this energy range 3 ways:

Antilogs (numbers): from [   ] to [   ]

Logs (exponents): from [   ] to [   ]

Orders of Magnitude: more than [   ] OM's.

NOTES (this page)

5a. Students should read this compressed antilog scale directly, then fill in the numbers (antilogs). They don't need their *Log Tapes* yet.

5b. Students should read this compressed log scale directly, then consult their *Log Tapes* for accuracy to 3 decimal places.

Ask volunteers to explain the relationship between these 2 columns of numbers. (Each antilog on the left equals each log on the right. Example: $7 = 10^{0.845}$.)

GLASToids!

8. GLAST will detect gamma rays energies from 10 million to 300 billion electron volts

Antilog range: from 10,000,000 to 300,000,000,000

Log range: from 7.000 to 11.477

Orders of Magnitude: more than 4 OM's

ANSWERS (this page)

5a.		5b.	
	8,000		3.903
	3,000		3.477
	600		2.778
	400		2.602
	90		1.954
	50		1.699
	10		1.000
	7		0.845
	.5		1.699
	.2		1.301
	.07		2.845
	.01		2.000
	.009		3.954
	.003		3.477
	.0008		4.903
	.0002		4.301

5c. The *Log Tape* contains *4 repeating cycles*.

5d. As the log of a number gets "very" negative, *the number itself gets very close to zero*.

5e. If a number is greater than zero but less than one, *the log of this number is negative*.

5f.
$\overline{1}.699 = -0.301$
$\overline{1}.301 = -0.699$
$\overline{2}.778 = -1.222$
$\overline{2}.477 = -1.523$

6.

antilog range	log range
1 to 10:	0 to 1
10 to 100:	1 to 2
1 to .1:	0 to -1
.1 to .01:	-1 to -2

7.

log range	antilog range
-3 to -4:	.001 to .0001
-4 to -5:	.0001 to .00001
3 to 4:	1,000 to 10,000
4 to 5:	10,000 to 100,000

NOTES: **Activity B1**

STRING CALCULATOR

Activity B2

1. Cut a piece of string as long as *half* of your *Log Tape*, plus an extra three finger-widths to tie a knot at each end.

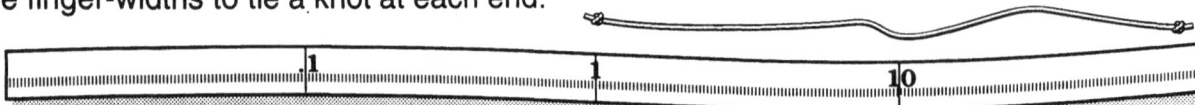

2a. Add these **logs** to find a total. Then fill in the **number**.

log 2 = **0.301**
+ log 3 =
―――――
log ___ =
 ↗   ↗
Number log N

2b. Then confirm these totals by measuring the first log distance with string, and adding it to the second log distance on your tape.

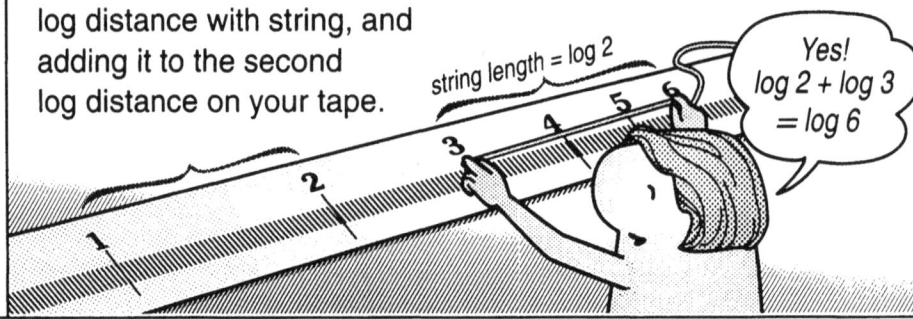

string length = log 2

Yes! log 2 + log 3 = log 6

3. For each problem below:
➤ Do a string calculation to fill in the *smallest* grey box.
➤ Record and add **logs** in the *larger* boxes.
➤ Confirm that **log M + log N = log MN**.

log 4 =	log 5 =	log 6 =	log 10 =
+ log 5 =	+ log 6 =	+ log 6 =	+ log 6 =
log **20** =	log ___ =	log ___ =	log ___ =
log 4 + log 5 = log 4x5			

log 10 =	log 2 =	log 16 =	INVENT:
+ log 10 =	+ log 12 =	+ log 1.5 =	
log ___ =	log ___ =	log ___ =	

4. Show that **log M + log N = log MN** for positive numbers less than one. (Start at log 1 = 0 as before, but extend the string LEFT.

log .2 =	log .4 = $\overline{1}.602$	log 1/2 =	log 1/10 =
+ log .3 =	+ log .04 = $\overline{2}.602$ positive!	+ log 1/5 =	+ log 1/10 =
log ___ =	log ___ =	log ___ =	log ___ =

5. Find inverse logs: Equal but opposite string lengths add to zero, while their corresponding numbers multiply to one! Show that **log M + log N = log 1 = 0**

log .05 =	INVENT:	INVENT:	INVENT:
+ log 20 =			
log ___ =			
log .05 + log 20 =			
log .05 x 20 = log 1 = 0			

continue

OVERVIEW / OBJECTIVES

Students will add and subtract log distances on their *Log Tapes* to discover that the corresponding numbers multiply and divide. This will lead them to an experiential understanding of the laws of logarithms.

TIME: 1 hour and 15 minutes.

INTRODUCTION

1. Stick a *Log Tape* to your board for students to see. Label these log positions on the positive (right) half of the tape, using vertical lines like these:

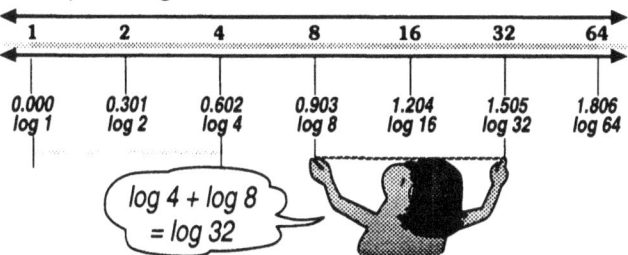

a. Measure any log distance in string *(say 0.602)* and add it to any other log distance *(0.903)* to get a total *(1.505)*. What happens to the antilogs? *(They always multiply.)* Have students demonstrate other examples. In general, **log M + log N = log MN**.

b. Next try some subtraction problems. The illustration above shows equally well that log 32 - log 4 = log 8. In general, **log M - log N = log M/N**

NOTES (this page)

1. This string is like temporary memory in a computer. It stores *one* of the log distances, which then can be added to (or subtracted from) any other log distance on the *Log Tape* itself.

ANSWERS (this page)

2a.
log 2 = 0.301
+ log 3 = 0.477
log 6 = 0.778

2b. Students should confirm by measuring: log 2 + log 3 = log 6
0.301 + 0.477 = 0.778

3. Added logs multiply numbers (N's greater than 1).

log 4 = **0.602**	log 5 = **0.699**	log 6 = **0.778**	log 10 = **1.000**
+ log 5 = **0.699**	+ log 6 = **0.778**	+ log 6 = **0.778**	+ log 6 = **0.778**
log 20 = **1.301**	log 30 = **1.477**	log 36 = **1.556**	log 60 = **1.778**
log 4 + log 5 = log 4×5	log 5 + log 6 = log 5×6	log 6 + log 6 = log 6×6	log 10 + log 6 = log 10×6
log 10 = **1.000**	log 2 = **0.301**	log 16 = **1.204**	
+ log 10 = **1.000**	+ log 12 = **1.079**	+ log 1.5 = **0.176**	INVENT:
log 100 = **2.000**	log 24 = **1.380**	log 24 = **1.380**	
log 10 + log 10 = log 10×10	log 2 + log 12 = log 2×12	log 16 + log 1.5 = log 16×1.5	

4. Added logs multiply numbers (positive N's less than 1).

log .2 = $\bar{1}$.301	log .4 = $\bar{1}$.602	log 1/2 = $\bar{1}$.699	log 1/10 = $\bar{1}$.000
+ log .3 = $\bar{1}$.477	+ log .04 = $\bar{2}$.602	+ log 1/5 = $\bar{1}$.301	+ log 1/10 = $\bar{1}$.000
log .06 = $\bar{2}$.778	log .016 = $\bar{2}$.204	log 1/10 = $\bar{1}$.000	log 1/100 = $\bar{2}$.000
log .2 + log .3 = log .2×.3	log .4 + log .04 = log .4×.04	log 1/2 + log 1/5 = log 1/2×1/5	log 1/2 + log 1/5 = log 1/2×1/5

5. Added logs multiply numbers (inverse N's).

log .05 = $\bar{2}$.699	
+ log 20 = **1.301**	
log 1 = **0.000**	INVENT:
log .05 + log 20 = log .05×20 = log 1 = 0	

ANSWERS (next page)

6. Added logs multiply numbers (3 N's).

log 2 = **0.301**	log .3 = $\bar{1}$.477	log 1/2 = $\bar{1}$.699	
+ log 3 = **0.477**	+ log .4 = $\bar{1}$.602	+ log 1/2 = $\bar{1}$.699	INVENT:
+ log 4 = **0.602**	+ log .5 = $\bar{1}$.699	+ log 4 = **0.602**	
log 24 = **1.380**	log .06 = $\bar{2}$.778	log 1 = **0.000**	
log 2 + log 3 + log 4 = log 2×3×4	log .3 + log .4 + log .5 = log .2×.3×.4	log 1/2 + log 1/2 + log 4 = log 1/2×1/2×4 = log 1 = 0	

7. Subtracted logs divide numbers (N's greater than 1).

log 6 = **0.778**	(+1) log 3 = ①.**477**	log 60 = **1.778**	(+1) log 12 = ②.**079**
- log 3 = **0.477**	- log 6 = **0.778**	- log 12 = **1.079**	- log 60 = **1.778**
log 2 = **0.301**	(-1) log 1/2 = $\bar{1}$.**699**	log 5 = **0.699**	(-1) log .2 = $\bar{1}$.**301**
log 6 - log 3 = log 6/3	log 3 - log 6 = log 3/6	log 60 - log 12 = log 60/12	log 12 - log 60 = log 12/60

To keep the mantissa positive, 1 is "borrowed" (added) on the top line, then "repaid" (subtracted) on the bottom line.

8. Multiply the log to raise its number to that power.

log 4 = **0.602**	log 2 = **0.301**	log .5 = $\bar{1}$.**699**	log .1 = $\bar{1}$.**000**
× 3 × 3	× 5 × 5	× 2 × 2	× 4 × 4
log **64** = **1.806**	log **32** = **1.505**	log **.25** = $\bar{1}$.**398**	log **.0001** = $\bar{4}$.**000**
3(log 4) = log 4^3	5(log 2) = log 2^5	2(log .5) = log $.5^2$	4(log .1) = log $.1^4$

9. Divided the log to take that root of its number.

log 64 = **1.806**	log 64 = **1.806**	log 64 = **1.806**	log .16 = $\bar{2}$ = 1.204
÷ 2 ÷ 2	÷ 3 ÷ 3	÷ 6 ÷ 6	÷ 2 ÷ 2
log **8** = **0.903**	log **4** = **0.602**	log **2** = **0.301**	log **.4** = $\bar{1}$.**602**
log 64 / 2 = log $\sqrt[2]{64}$	log 64 / 3 = log $\sqrt[3]{64}$	log 64 / 6 = log $\sqrt[6]{64}$	log .16 / 2 = log $\sqrt[2]{.16}$

10. To multiply numbers, **add** their logs. To divide numbers, **subtract** their logs. To raise a number to a power, **multiply** its log by that power. To take the root of a number, **divide** its log by that root.

GLASToids!

11.

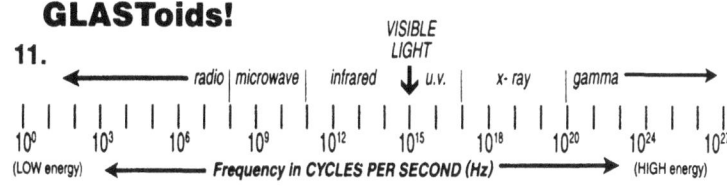

MATERIALS (additional)

☐ String. Precut to 60 cm lengths, or supply scissors and a meter stick.

NOTES: **Activity B2**

STRING CALCULATOR

Activity B2

6. Use string to show that **log L + log M + log N = log LMN**.

| log 2 =
 + log 3 =
 + log 4 =
 log   =

 log 2 + log 3 + log 4 =
 log 2 x 3 x 4 | log .3 =
 + log .4 =
 + log .5 =
 log   = | log 1/2 =
 + log 1/2 =
 + log 4 =
 log   = | INVENT: |

7. Show that **log M − log N = log M/N**. (Hint: measure the *difference* in string. If *positive*, apply to the *right* of log 1 = 0; if *negative*, apply to the *left*.)

| log 6 =
 − log 3 =
 log   = | log 3 = $\bar{1}$.477 (add 1)
 − log 6 = − 0.778 <br> log   = $\bar{1}$.699 (sub 1) | log 60 =
 − log 12 =
 log   = | log 12 =
 − log 60 =
 log   = |

8. Lay the correct string length end to end, the given number of times. Do this first! Then show that **x (log N) = log N^x**.

| log 4 =
 x 3   x 3 <br> log   = | log 2 =
 x 5   x 5 <br> log   = | log .5 =
 x 2   x 2 <br> log   = | log .1 =
 x 4   x 4 <br> log   = |

9. Fold (or imagine dividing) the correct string length into equal pieces. Do this first! Then show that **(log N) / x = log $\sqrt[x]{N}$**

| log 64 =
 ÷ 2   ÷ 2 <br> log   = | log 64 =
 ÷ 3   ÷ 3 <br> log   = | log 64 =
 ÷ 6   ÷ 6 <br> log   = | log .16 = $\bar{2}$ + 1.204
 ÷ 2   ÷ 2 <br> log   = |

10. To **multiply** numbers, _____ their logs. To **divide** numbers, _____ their logs. To **raise** a number **to a power**, _____ its log by that power. To **take the root** of a number, _____ its log by that root.

11. So, find that page of *GLASToids* again:
The electromagnetic (EM) spectrum is commonly grouped into 7 sections, spanning many orders of magnitude. Label these on this log ruler.

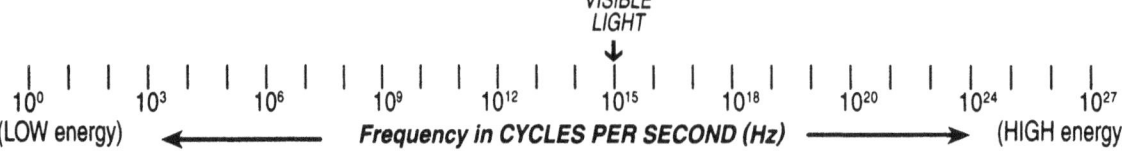

LOG TAPE

Supplement to Activity B1

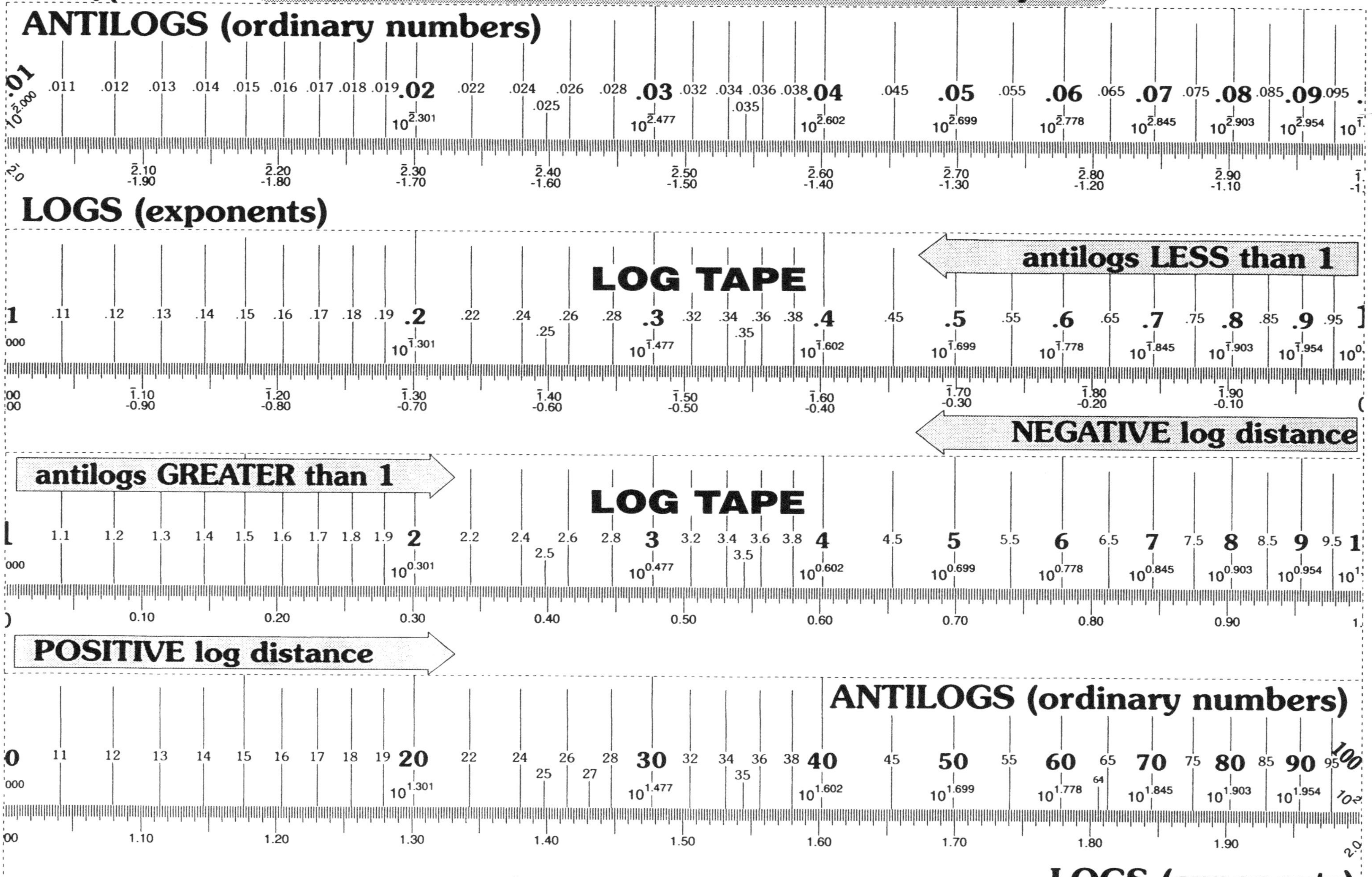

ANTILOGS (ordinary numbers)

LOGS (exponents)

LOG TAPE — antilogs LESS than 1

NEGATIVE log distance

LOG TAPE — antilogs GREATER than 1

POSITIVE log distance

ANTILOGS (ordinary numbers)

LOGS (exponents)

PAGE 31

Copyright © 2003 by TOPS Learning Systems, Canby OR 97013.

NO-STRINGS CALCULATING

Activity B3
1 of 1 page

1. Use your *Log Tape* to complete this "Look-Up Table."

Use these data to work the problems below.

antilog	.2	.5			3	.4			8
log	$\overline{1}.301$		0	0.301		0.699	0.778		

antilog		10	12	16	18			36	
log	0.954					1.301	1.398		2.301

2a. State the log rule. Then evaluate each section with your Look-Up table.

FORMULA: log M + log N = log MN.
CONCEPT: **Add** logs to **multiply** numbers.

| 9 × 4 = _____ 0.954 + _____ | 10 × 20 = | .2 × 10 = | 36 × ½ = |

2b. FORMULA: _____
CONCEPT: _____

| 36 / 3 = | 200 ÷ 1 = | 20 / 5 = | 4 ÷ 20 = |

2c. FORMULA: _____
CONCEPT: _____

| 2^3 = | 4^2 = | 5^2 = | 1^5 = |

2d. FORMULA: _____
CONCEPT: _____

| $4^{1/2}$ = | $\sqrt[2]{9}$ = | $\sqrt[4]{16}$ = | $36^{1/2}$ = |

3. You know what to do!

Glast astronomers love to talk about gamma photon energies in terms of "keV's, MeV's, GeV's and TeV's." Define these energies in terms of electron volts (eV's). (Photons of visible light typically have energies of about 2 electron volts or 2 eV).

| 1 keV = 10^3 eV | 1 MeV = | _____ eV | _____ eV |

OVERVIEW / OBJECTIVES

Students will first develop a simplified log table using information on their *Log Tapes*. Then they will use it to solve arithmetic problems by looking up and combining logs, and finding the antilog. Because these problems are extremely simple, students will appreciate the logic of logarithms without getting bogged down in the arithmetic detail and error.

TIME: 20 minutes.

NOTES (this page)

1-2. Students should use their *Log Tapes* to develop the Look-Up Table in step 1, then put them away. In step 2 they should rely *exclusively* on their Look-Up Table to solve each problem.

Tables like these were how mathematicians, scientists and engineers solved problems 300 years ago, before the invention of slide rules. The problems, of course, were more complex, and the log tables were more accurate and extensive. But the logic and methods are identical.

ANSWERS (this page)

1. See table below.

2a. FORMULA: $\log M + \log N = \log MN$
CONCEPT: **Add logs to multiply numbers.**

9 × 4 = **36**	10 × 20 = **200**	.2 × 10 = **2**	36 × ½ = **18**
0.954	1.000	$\overline{1}.301$	1.556
+ 0.602	+ 1.301	+ 1.000	+ $\overline{1}.699$
1.556	2.301	0.301	1.257

2b. FORMULA: $\log M - \log N = \log M/N$
CONCEPT: **Subtract logs to divide numbers.**

36/3 = **12**	200 ÷ 1 = **200**	20/5 = **4**	4 ÷ 20 = **.2**
1.556	2.301	1.301	0.602
- 0.477	- 0.000	- 0.699	- 1.301
1.079	2.301	0.602	$\overline{1}.301$

2c. FORMULA: $n \log M = \log M^n$
CONCEPT: **Multiply logs to raise a number by that power.**

$2^3 = 8$	$4^2 = 16$	$5^2 = 25$	$1^5 = 1$
0.301	0.602	0.699	0.000
× 3	× 2	× 2	× 5
0.903	1.204	1.398	0.000

2d. FORMULA: $\log M / n = \log \sqrt[n]{M}$
CONCEPT: **Divide logs to take that root of a number**

$4^{1/2} = 2$	$\sqrt[2]{9} = 3$	$\sqrt[4]{16} = 2$	$36^{1/2} = 6$
0.602 / 2 = 0.301	0.954 / 2 = 0.477	1.204 / 4 = 0.301	1.556 / 2 = 0.778

GLASToids!

3.

| 1 keV = 10^3 eV | 1 MeV = 10^6 eV | 1 GeV = 10^9 eV | 1 TeV = 10^{12} eV |

1. Look-Up Table:

antilog	.2	.5	**1**	**2**	3	4	**5**	**6**	8
log	$\overline{1}.301$	$\overline{1}.699$	0	0.301	**0.477**	**0.602**	0.699	0.778	**0.903**
antilog	**9**	10	12	16	18	**20**	**25**	36	**200**
log	0.954	**1.000**	1.079	1.204	**1.255**	1.301	1.398	**1.556**	2.301

NOTES: **Activity B3**

BASE-TWO SLIDE RULE

Activity C1
1 of 2 pages

1. Get the *Base-Two Slide Rule* supplement, and fold as usual. *Stop here*

2. Fit it with a hairline as you did before.

3. Draw *circles* around the tiny exponents (logs), and *boxes* around the larger numbers (antilogs), for the first seven calibrations as shown.

4. Show how **base 2** relates each circle to its box: ___ ___ ___ $2^3=8$ $2^4=16$ ___ ___

5. Add and subtract these **logs** (tiny circled exponents)... ...to multiply and divide these **antilogs** (large boxed numbers).

Hairline continues + and ×
C-scale continues − and ÷

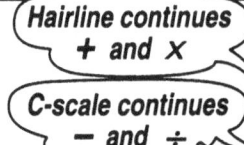

⑤+⑤ = ____. Thus, ... 32 × 32 = _____

⑥-④+③ = ____. Thus, ... 64 ÷ 16 × 8 = _____

④+④+④+④ = ____. Thus, 16 × 16 × 16 × 16 = _____ . 16^4 = _____ .

㉑-⑦-⑦-⑦ = ____. Thus, 2,097,152 ÷ 128 ÷ 128 ÷ 128 = ____. $\sqrt[3]{2{,}097{,}152}$ = ____.

6. Add and subtract **logs** (exponents) to multiply and divide **antilogs** (numbers).

$\log_2 8 + \log_2 8$ = __log$_2$ 64__. Thus, __8 × 8 = 64__.

$\log_2 4 + \log_2 32$ = _____. Thus, _____.

$\log_2 256 - \log_2 32$ = _____. Thus, _____.

$\log_2 2{,}048 - \log_2 512$ = _____. Thus, _____.

$\log_2 16 + \log_2 16 - \log_2 2$ = _____. Thus, _____.

$\log_2 512 - \log_2 64 + \log_2 16$ = _____. Thus, _____.

$3 \log_2 8 = \log_2 8 + \log_2 8 + \log_2 8$ = _____. Thus, 8^3 = _____.

$3 \log_2 32$ = _____ + _____ + _____ = _____. Thus, _____.

$\log_2 16{,}384 - \log_2 128 - \log_2 128 = \log_2$ _____ = _____. Thus, $\sqrt[?]{16{,}384}$ = _____.

$\log_2 256 -$ _____ − _____ − _____ − _____ = _____ = ___. Thus, _____.

7. Multiply and divide **antilogs** (numbers) to add and subtract **logs** (exponents).

4 × 8 × 16 = _____. Thus, _____.

8,192 ÷ 128 ÷ 8 = _____. Thus, _____.

16^4 = _____. Thus, $4 \log_2 16$ = _____ + _____ + _____ + _____.

$\sqrt[3]{512}$ = _____. Thus, $\log_2 512 -$ _____ − _____ − _____ = _____ = 0.

8. Explain why a slide rule works. How does it multiply and divide numbers?

continue

OVERVIEW / OBJECTIVES

Students will construct *Base-Two Slide Rules* that add and subtract base-2 exponents (log distances), in order to multiply and divide corresponding powers of two. They'll use these slide rules to generate both log and antilog equations, learning to translate one in terms of the other.

TIME: 50 minutes.

INTRODUCTION

Draw a line on your board, and label *exponents* 0 to 4 under the line.

(a) In base 10, what *numbers* would we write above these exponents? *(1, 10, 100, 1000, 10000).*

Let's not do that. Instead, we'll write numbers corresponding to *base 2*. If each of these exponents represents a power of 2, what numbers would we write above them?

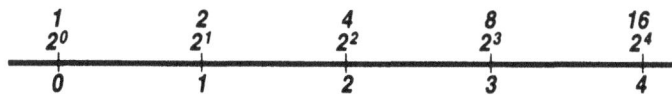

(b) Since the log of a *number* is always an *exponent*, we can add $\log_2 N$ labels under each exponent. And because the antilog of an *exponent* is always a *number*, we can add $\text{antilog}_2 \text{ exp}$ labels over each pair of numbers.

(antilog$_2$ N = exp)

antilog$_2$ 0	antilog$_2$ 1	antilog$_2$ 2	antilog$_2$ 3	antilog$_2$ 4
1	2	4	8	16
2^0	2^1	2^2	2^3	2^4
0	1	2	3	4
log$_2$ 1	log$_2$ 2	log$_2$ 4	log$_2$ 8	log$_2$ 16

(log$_2$ N = exp)

Notice the use of subscript "2". *(By convention, if no subscript is written, it is assumed to be base 10, which this isn't!)*

(c) In this lesson you will construct base-2 slide rules with exponent scales like these. As you add and subtract exponents on this scale, what will happen to the corresponding numbers? *(They'll multiply and divide.)*

(d) What happens when we extend this scale leftward, below zero, into negative exponents? Who can extend the scale?

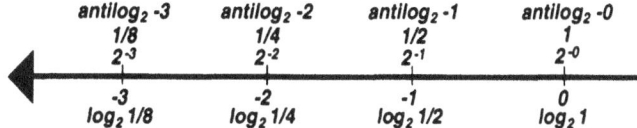

NOTES (this page)

1-2. This slide rule folds together exactly like the previous two. Students add hairlines constructed in activity A1.

3. Exponents standing alone, without their base, look like ordinary numbers unless linked by equal signs to their logs. In this step students circle exponents and box ordinary numbers to keep this distinction firmly in mind.

5. This *Base-Two Slide Rule* is a log scale! So where are the logs? All those tiny base-two exponents, (some circled) are logs. It is probably impossible to say this too many times: **logs are exponents!** For simplicity, the log calibrations read 0, 1, 2, 3.... They could just as well read $\log_2 1$, $\log_2 2$, $\log_2 4$, $\log_2 8$ ...

ANSWERS (this page)

5. $5 + 5 =$ *10*. Thus, $32 \times 32 =$ *1,024*
$6 - 4 + 3 =$ *5*. Thus, $64 \div 16 \times 8 =$ *32*
$4 + 4 + 4 + 4 =$ *16*. Thus, $16 \times 16 \times 16 \times 16 =$ *65,536*
$\qquad\qquad\qquad\qquad\qquad 16^4 =$ *64*
$21 - 7 - 7 - 7 =$ *0*. Thus, $2,097,152 \div 128 \div 128 \div 128 =$ *0*
$\qquad\qquad\qquad\qquad\qquad \sqrt[3]{2,097,152} =$ *128*

6. $\log_2 8 + \log_2 8 =$ *$\log_2 64$*. Thus, *$8 \times 8 = 64$*
$\log_2 4 + \log_2 32 =$ *$\log_2 128$*. Thus, *$4 \times 32 = 128$*
$\log_2 256 - \log_2 32 =$ *$\log_2 8$*. Thus, *$256 \div 32 = 8$*
$\log_2 2,048 - \log_2 512 =$ *$\log_2 4$*. Thus, *$2,048 \div 512 = 4$*
$\log_2 16 + \log_2 16 - \log_2 2 =$ *$\log_2 128$*. Thus, *$16 \times 16 \div 2 = 128$*
$\log_2 512 - \log_2 64 + \log_2 16 =$ *$\log_2 128$*.
$\qquad\qquad\qquad$ Thus, *$512 \div 64 \times 16 = 128$*
$3 \log_2 8 = \log_2 8 + \log_2 8 + \log_2 8 =$ *$\log_2 512$*.
$\qquad\qquad\qquad$ Thus, $8^3 =$ *512*
$3 \log_2 32 =$ *$\log_2 32$* + *$\log_2 32$* + *$\log_2 32$* = *$\log_2 32,768$*
$\qquad\qquad\qquad$ Thus, $32^3 =$ *32,768*
$\log_2 16,384 - \log_2 128 - \log_2 128 =$ *$\log_2 1 = 0$*.
$\qquad\qquad\qquad$ Thus, $\sqrt[2]{16,384} =$ *128*
$\log_2 256 -$ *$\log_2 4$* $-$ *$\log_2 4$* $-$ *$\log_2 4$* $-$ *$\log_2 4$* $=$ *$\log_2 1 = 0$*.
$\qquad\qquad\qquad$ Thus, $\sqrt[4]{256} =$ *4*

7. $4 \times 8 \times 16 =$ *512*. Thus, *$\log_2 4$* $+$ *$\log_2 8$* $+$ *$\log_2 16$* $=$ *$\log_2 512$*
$8,192 \div 128 \div 8 =$ *8*.
$\qquad\qquad$ Thus, *$\log_2 8,192$* $-$ *$\log_2 128$* $-$ *$\log_2 8$* $=$ *$\log_2 8$*
$16^4 =$ *65,536*. Thus, $4 \log_2 16 =$
$\qquad$ *$\log_2 16$* $+$ *$\log_2 16$* $+$ *$\log_2 16$* $+$ *$\log_2 16$* $=$ *$\log_2 65,536$*
$\sqrt[3]{512} =$ *8*. Thus, $\log_2 512 -$ *$\log_2 8$* $-$ *$\log_2 8$* $-$ *$\log_2 8$* $=$ *$\log_2 1 = 0$*

8. A slide rule adds and subtracts logs to multiply and divide numbers.

BASE-TWO SLIDE RULE

Activity C1
2 of 2 pages

9. Complete this base-two table. (To maximize learning, please work **across** the table, not down.)

A. EXPONENTS ($2^{exp} = N$)	B. ANTILOGS (antilog$_2$ exp = N)	C. LOGS (log$_2$ N = exp)
$2^7 = 128$	antilog$_2$ 7 = 128	log$_2$ 128 = 7
$2^6 =$		
= 32		
	antilog$_2$ 4 =	
		log$_2$ 8 =
		= 2
	= 2	
= 1		
$2^{-1} =$		
	antilog$_2$ -2 =	
		log$_2$ 1/8 =
	= 1/16	
		= -5

10. In **base-two**...

a. Antilogs may also be expressed as _____ and _____.

b. Logs may also be expressed as _____,

c. The antilog of the log of a number is _____.

d. The log of the antilog of an exponent is _____.

e. As log$_2$ N gets "very" negative, what happens to **N**? _____.

f. If log$_2$ N is greater than 0, what can you say about N? _____.

11a. Get your **Adding** Slide Rule. Renumber the first 9 divisions to create a **base-3** multiplier!

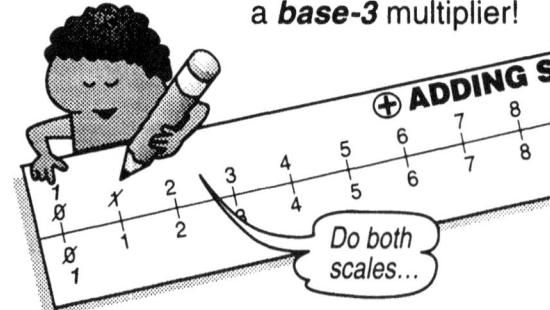

Do both scales...

11b. Invent a sample problem:

11c. How does this renumbered slide rule work?

12a. GLAST will absorb gamma photons using 19 layers of dense interactive material: a gamma photon hits the **1st** layer and turns into **2** particles (electron and positron); these hit the **2nd** layer to become **4** particles, which hit the **3rd** layer to become **8** particles, which hit the **4th** layer to become _____ particles, and so on.

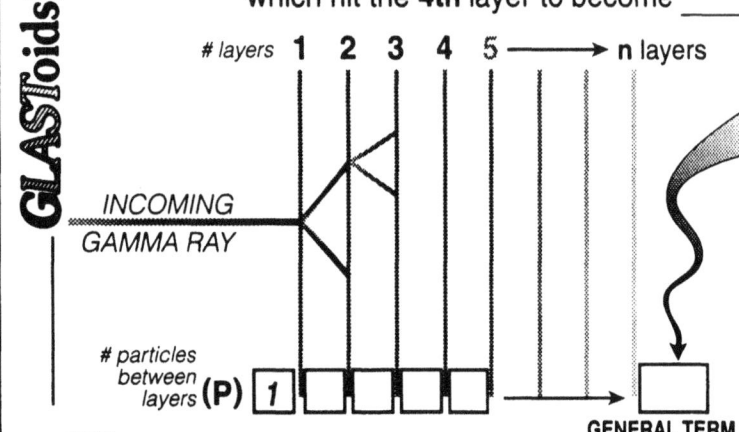

b. Draw the "pair production" of particles passing through the first 4 layers. Record the number of particles between layers in each box, and write the general term.

c. If a gamma ray is too energetic for GLAST to fully absorb, how many particles will be produced at the 19th layer before they exit the telescope?

NOTES (this page)

11. If you directed your class to skip STRAND A, students can substitute their *Base-Two Slide Rule* for their missing *Adding Slide Rule*. Suggest they write lightly, so that after they demonstrate their solutions to you, they can erase these numbers. They will still need to use their base-2 rules again, in the remainder of STRAND C and in activity E2.

ANSWERS (this page)

9.

A EXPONENTS	B ANTILOGS	C LOGS
$2^7 = 128$	$antilog_2\ 7 = 128$	$log_2\ 128 = 7$
$2^6 = 64$	$antilog_2\ 6 = 64$	$log_2\ 64 = 6$
$2^5 = 32$	$antilog_2\ 5 = 32$	$log_2\ 32 = 5$
$2^4 = 16$	$antilog_2\ 4 = 16$	$log_2\ 16 = 4$
$2^3 = 8$	$antilog_2\ 3 = 8$	$log_2\ 8 = 3$
$2^2 = 4$	$antilog_2\ 2 = 4$	$log_2\ 4 = 2$
$2^1 = 2$	$antilog_2\ 1 = 2$	$log_2\ 2 = 1$
$2^0 = 1$	$antilog_2\ 0 = 1$	$log_2\ 1 = 0$
$2^{-1} = 1/2$	$antilog_2\ -1 = 1/2$	$log_2\ 1/2 = -1$
$2^{-2} = 1/4$	$antilog_2\ -2 = 1/4$	$log_2\ 1/4 = -2$
$2^{-3} = 1/8$	$antilog_2\ -3 = 1/8$	$log_2\ 1/8 = -3$
$2^{-4} = 1/16$	$antilog_2\ -4 = 1/16$	$log_2\ 1/16 = -4$
$2^{-5} = 1/32$	$antilog_2\ -5 = 1/32$	$log_2\ 1/32 = -5$

10. In base-two,
   a. Antilogs may also be expressed as *ordinary numbers*, and *powers of two*.
   b. Logs may also be expressed as *exponents*.
   c. The antilog of the log of a number is *a number*.
   d. The log of the antilog of an exponent is *an exponent*.
   e. As $log_2\ N$ gets "very" negative, *N gets very close to 0*.
   f. If log N is greater than 0, *then N is greater than 1*.

11a. This step directs students to the **Adding** *Slide Rule* (not the *Base-2 Slide Rule*). Numbers 0-8 should be changed to read *1, 3, 9, 27, 81, 243, 729, 2187, 6561.*

11b. A sample problem (one of many) that can be calculated on this renumbered slide rule: *9 x 27 = 243.*

11c. *This renumbered slide rule adds and subtracts base three logs to multiply and divide base three numbers.*

GLAStoid!

12a. ...*16* particles, and so on.

12b.

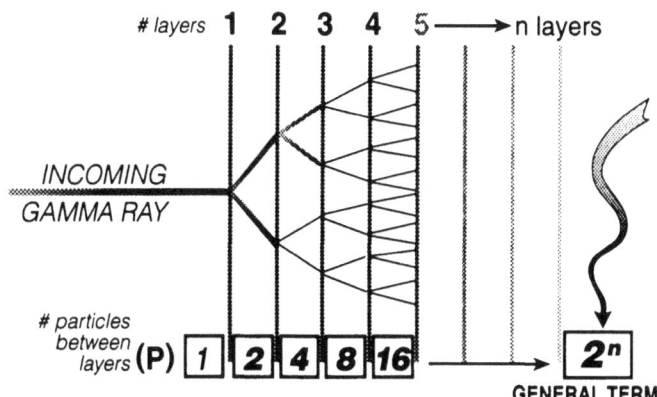

12c. $2^n = 2^{19} = 524,288$ particles

NOTES: **Activity C1**

EXPONENTIAL UPS & DOWNS

1. Let's summarize the relationship between logs (exponents) and ordinary numbers:

The log of a number is an exponent.

The exponent of a base is a number.

These actually say the same thing.

The antilog of an exponent is a number.

$B^{exp} = N$ (RAISED, antilog)

$\log_B N = exp$ (PULLED DOWN, log) (N>0)

$\text{antilog}_B exp = N$

2. Practice "raising" and "pulling down" base-2 exponents. See your **Base-2 Slide Rule**.

$2^3 = $ ___	$\log_2$ ___ $= 3$
	$\log_2$ ___ $= 5$
$2^{11} = $ ___	
	$\log_2 8{,}192 = $ ___

	$\log_2 512 = $ ___
$2^{-} = 65{,}536$	
	$\log_2$ ___ $= 6$
$2^{-} = 1$	

3. Practice raising and pulling down base-10 exponents. See your **Log Tape**.

$10^2 = $ ___	
	$\log$ ___ $= -5$
$10^{1.699} = $ ___	
	$\log 2{,}000 = $ ___

	$\log \frac{1}{10} = $ ___
$10^{-} = 32$	
	$\log$ ___ $= 0$
$10^{-} = \frac{1}{100}$	

4. Practice raising and pulling down exponents in **ANY base**! Use a calculator.

$5^4 = $ ___	
	$\log_7$ ___ $= 3$
$3^4 = $ ___	
	$\log_8 \frac{1}{64} = $ ___

	$\log_6 216 = $ ___
$12^{-} = 144$	
	$\log_{17}$ ___ $= 1$
$(\sqrt{3})^{-} = 1$	

5. Evaluate these logs and antilogs. (Hint: First convert to $B^{exp} = N$ inside each box.)

$4^3 = 64$	$\text{antilog}_4 3 = $ ___	$\log_4$ ___ $= 3$
	antilog_3	$\log_3 \frac{1}{9} = $ ___
	$\text{antilog}_{\_} 3 = 1331$	
	antilog_7	$\log_7 1 = $ ___

antilog_3	$\log_{\_} 9 = 2$
$\text{antilog}_{\_} -3 = \frac{1}{64}$	
antilog	$\log_{\_} 243 = 5$
antilog_{100} ___ $= 100$	

GLASToids!

6a. Examine your slide rule to discover an interesting regularity.

1 thousand	1 million	1 billion	1 trillion
10^3	10^6		
$\approx 2^{10}$			

6b. Express the projected cost of GLAST in base-2 dollars and base-10 dollars.

GLAST budget (in base 2)	= ___ dollars
GLAST budget (in base 10)	= ___ dollars

OVERVIEW / OBJECTIVES

Students will use their *Base-Two Slide Rules*, *Log Tapes*, and calculators to practice "raising" exponents in base notation, and "pulling down" exponents in log notation, in different bases. They'll come to appreciate that antilog notation expresses exactly the same idea as raising a base to a power.

TIME: 20 minutes.

INTRODUCTION

★ Any positive number (**N**) can also be written as a base (**B**) raised to some power or exponent (**exp**). If N = 64, for example, we can express this number using different bases and exponents. List some possibilities in a number table as below.

Antilogs are numbers too! List some of these on the right of the same table, inviting volunteers to help. Reading left to right, notice that both equations specify base, exponent and number in the same order.

Logs are exponents! These don't belong in the number table because they're not ordinary numbers. Ask volunteers to list these in an separate exponent table. Notice how taking the log of a number "pulls down the exponent."

Numbers		Exponents
$B^{exp} = N$	$antilog_B\ exp = N$	$log_B\ N = exp$
$8^2 = 64$	$antilog_8\ 2 = 64$	$log_8\ 64 = 2$
$4^3 = 64$	$antilog_4\ 3 = 64$	$log_4\ 64 = 3$
$2^6 = 64$	$antilog_2\ 6 = 64$	$log_2\ 64 = 6$
$10^{1.806} = 64$	$antilog\ 1.806 = 64$	$log\ 64 = 1.806$
$4096^{1/2} = 64$	$antilog_{4096}\ 1/2 = 64$	$log_{4096}\ 64 = 1/2$

This table summarizes three basic ways that exponents relate to numbers. Become fluent with these relationships, and you've mastered the language of logarithms!

NOTES (this page)

6b. The projected cost of GLAST in base-2 dollars is hidden in the GLASToids. Finding this factoid, students can easily convert to base-10 dollars using the equivalence table developed in 6a.

ANSWERS (this page)

2. Using *Base-2 Slide Rule*:

$2^3 = 8$	$log_2\ 8 = 3$	$2^9 = 512$	$log_2\ 512 = 9$
$2^5 = 32$	$log_2\ 32 = 5$	$2^{16} = 65{,}536$	$log_2\ 65{,}536 = 16$
$2^{11} = 2048$	$log_2\ 2048 = 11$	$2^6 = 64$	$log_2\ 64 = 6$
$2^{13} = 8192$	$log_2\ 8192 = 13$	$2^0 = 1$	$log_2\ 1 = 0$

3. Using *Log Tape*:

$10^2 = 100$	$log\ 100 = 2$	$10^{-1} = 1/10$	$log\ 1/10 = -1$
$10^{-5} = .00001$	$log\ .00001 = -5$	$10^{1.505} = 32$	$log\ 32 = 1.505$
$10^{1.699} = 50$	$log\ 50 = 1.699$	$10^0 = 1$	$log\ 1 = 0$
$10^{3.301} = 2000$	$log\ 2000 = 3.301$	$10^{-2} = 1/100$	$log\ 1/100 = -2$

4. In any base:

$5^4 = 625$	$log_5\ 625 = 4$	$6^3 = 216$	$log_6\ 216 = 3$
$7^3 = 343$	$log_7\ 343 = 3$	$12^2 = 144$	$log_{12}\ 144 = 2$
$3^4 = 81$	$log_3\ 81 = 4$	$17^1 = 17$	$log_{17}\ 17 = 1$
$8^{-2} = 1/64$	$log_8\ 1/64 = -2$	$(\sqrt{3})^0 = 1$	$log_{\sqrt{3}}\ 1 = 0$

5. Evaluate logs & antilogs:

$4^3 = 64$	$antilog_4\ 3 = 64$	$log_4\ 64 = 3$
$3^{-2} = 1/9$	$antilog_3\ -2 = 1/9$	$log_3\ 1/9 = -2$
$11^3 = 1331$	$antilog_{11}\ 3 = 1331$	$log_{11}\ 1331 = 3$
$7^0 = 1$	$antilog_7\ 0 = 1$	$log_7\ 1 = 0$
$3^2 = 9$	$antilog_3\ 2 = 9$	$log_3\ 9 = 2$
$4^{-3} = 1/64$	$antilog_4\ -3 = 1/64$	$log_4\ 1/64 = -3$
$3^5 = 243$	$antilog_3\ 5 = 243$	$log_3\ 243 = 5$
$100^1 = 100$	$antilog_{100}\ 1 = 100$	$log_{100}\ 100 = 1$

GLASTOIDS

6a.

1 thousand	1 million	1 billion	1 trillion
10^3	10^6	10^9	10^{12}
$\approx 2^{10}$	$\approx 2^{20}$	$\approx 2^{30}$	$\approx 2^{40}$

6b.

GLAST budget (in base 2) = 400×2^{20} dollars

GLAST budget (in base 10) = 400×10^6 = 400,000,000 dollars

LOG ALGEBRA

Activity C3
1 of 2 pages

Use your *Log Tape* or *Base-2 Slide Rule* as a reference to solve these problems. Always begin by taking the log of both sides, then show *all* steps, *all* the way down.

1. Solve in base 10:
$$3^x = 9$$
take the log of both sides
$$\log 3^x = \log 9$$
apply rule: $\log N^x = x \log N$
("pull down" the exponent)
$$x \log 3 = \log 9$$
$$x = \frac{\log 9}{\log 3}$$
$$x = \frac{}{.477}$$
$$x =$$

2. Solve in base 10.
$$6^x = 36$$

3. Solve in base 10.
$$16^x = 4$$

4. Solve again in base 2.
$$16^x = 4$$
$$\log_2 16^x = \log_2 4$$

5. Solve in base 10.
$$3^x = 28$$

6. Solve in base 2.
$$4^x = 32$$

7. Solve in base 10.
$$4^x = 32$$

8. Solve in any base!
$$n^x = 1$$

9. Solve in any base!
$$n^x = n$$

10. Solve in base 2.
$$8^x = (32)(16)$$
$$\log 8^x = \log (32)(16)$$
apply rule: $\log MN = \log M + \log N$
$$\log 8^x = \log 32 + \log 16$$

11. Solve in base 2.
$$64^x = (32)(128)$$

12. Solve in base 10.
$$5^x = (5)(25)$$

PAGE 40 BASE-TWO SLIDE RULE Copyright © 2003 by TOPS Learning Systems, Canby OR 97013

OVERVIEW / OBJECTIVES

Students will solve exponential equations where the unknown is contained in the exponent. They'll learn that taking base-10 or base-2 logs "pulls down the exponent," allowing the unknown to be isolated and solved.

TIME: 40 minutes.

INTRODUCTION

★ Who can solve this equation for x?
$$3^x = 27$$
Yes, x = 3. Equations with exponents are easy to solve by inspection when they are obvious whole numbers. But what about exponents that don't come out even? Who can solve this equation for x?
$$3^x = 28$$
We know that x will be greater than 3, but by how much? To know precisely, let's "take the base-10 log of both sides."
$$3^x = 28$$
$$\log 3^x = \log 28$$
Notice that this equation no longer equates ordinary numbers. The rules of logs (exponents) now apply. Let's review what we know about basic log rules:

(1) Add to multiply: $\log M + \log N = \log MN$
(2) Subtract to divide: $\log M - \log N = \log M/N$
(3) Multiply to raise to a power: $x \log B = B^x$
(4) Divide to take the root: $\frac{\log B}{x} = \log \sqrt[x]{B}$

Applying rule (3) we can now "pull down the exponent" and solve for x.
$$\log 3^x = \log 28$$
$$x \log 3 = \log 28$$
$$x = \log 28 / \log 3$$
$$x = 1.447 / 0.477 \text{ (from Log Tape)}$$
$$x = 3.034$$

★ Recall what a base-2 log scale looks like.

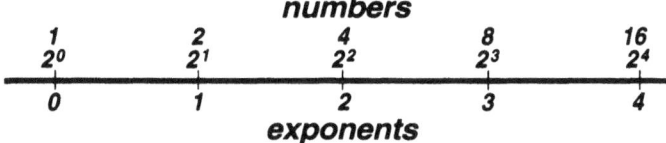

Where should we write 3 on the antilog scale above the line? Let's apply base-10 logs to determine where the exponent associated with 3 lies on the log scale under the line.
$$2^x = 3$$
$$\log 2^x = \log 3$$
$$x \log 2 = \log 3$$
$$x = \log 3 / \log 2$$
$$x = 0.447 / 0.301$$
$$x = 1.585$$

We see that 3 lies farther than halfway. Base-2 logs scales, like base-10, tend to bunch together in periodic cycles.

For extra credit or as a class project, you might ask students to develop a base-2 log scale that fixes all whole numbers (1-16) to a number line. If students do this accurately, they can slide two of these scales relative to each other in the manner of a slide rule, to multiply and divide integers.

ANSWERS (this page)

1. x = 2   2. x = 2   3. x = ½   4. x = ½
5. x = 3.304   6. x = 2½   7. x = 2.5   8. x = 0
9. x = 1
10. x = 3   11. x = 2   12. x = 3

NOTE: Selected problems from this answer key are detailed below to show typical steps students should follow:

3. Solve in base 10:
$$16^x = 4$$
$$\log 16^x = \log 4$$
$$x \log 16 = \log 4$$
$$x = \frac{\log 4}{\log 16}$$
$$x = \frac{0.602}{1.204}$$
$$x = \frac{1}{2}$$

4. Again, in base 2:
$$16^x = 4$$
$$\log_2 16^x = \log_2 4$$
$$x \log_2 16 = \log_2 4$$
$$x = \frac{\log_2 4}{\log_2 16}$$
$$x = \frac{2}{4}$$
$$x = \frac{1}{2}$$

8. Solve in any base:
$$n^x = 1$$
$$\log n^x = \log 1$$
$$x \log n = \log 1$$
$$x = \frac{\log 1}{\log n}$$
$$x = \frac{0}{\log n}$$
$$x = 0$$

ANSWERS (next page)

13. x = 4   14. x = ½   15. x = -2
16. x = 3   17. x = 5   18. x = 3
19. x = 4   20. x = 3

NOTE: Selected problems shown below:

14. Solve in base 2:
$$64^x = 512/64$$
$$\log_2 64^x = \log_2 \frac{512}{64}$$
$$x \log_2 64 = \log_2 512 - \log_2 64$$
$$x = \frac{\log_2 512 - \log_2 64}{\log_2 64}$$
$$x = \frac{9-6}{6} = \frac{3}{6}$$
$$x = \frac{1}{2}$$

18. Solve in base 10:
$$\sqrt[x]{27} = 3$$
$$\log \sqrt[x]{27} = \log 3$$
$$\frac{\log 27}{x} = \frac{\log 3}{1}$$
$$x \log 3 = \log 27$$
$$x = \frac{\log 27}{\log 3} = \frac{1.431}{.477}$$
$$x = 3$$

20. Solve in base 10:
$$6^x / 4 = (9)(6)$$
$$\log \frac{6^x}{4} = \log (9)(6)$$
$$x \log 6 - \log 4 = \log 9 + \log 6$$
$$x = \frac{\log 9 + \log 6 + \log 4}{\log 6}$$
$$x = \frac{.954 + .778 + .602}{.778}$$
$$x = 3$$

GLASTOIDS (next page)

21a. Solar mass: 2^{31} *suns*

21b. Solar mass in base 10: $2^{31} = 2 \times 2^{30} = 2 \times 10^9$ *suns*

NOTES: **Activity C3**

LOG ALGEBRA

Activity C3
2 of 2 pages

13. Solve in base 2.

$2^x / 8 = 2$

$\log 2^x / 8 = \log 2$

apply rule: log M/N = log M - log N

$\log 2^x - \log 8 = \log 2$

14. Solve in base 2.

$64^x = 512/64$

15. Solve in base 10.

$5^x = 1/25$

16. Solve in base 2.

$\sqrt[x]{64} = 4$

$\log \sqrt[x]{64} = \log 4$

apply rule: $\log \sqrt[x]{N} = \dfrac{\log N}{x}$

$\dfrac{\log 64}{x} = \dfrac{\log 4}{1}$

cross multiply:

$x \log 4 = \log 64$

17. Solve in base 2.

$\sqrt[x]{1024} = 4$

18. Solve in base 10.

$\sqrt[x]{27} = 3$

19. Solve in base 2.

$(2^x)(2^{2x}) = (64)(64)$

20. Solve in base 10.

$6^x / 4 = (9)(6)$

21. Galaxy 3C273, about 2.5 billion light years from earth, is an extremely "bright" source of gamma rays because it harbors a *really huge* black hole at its center.

a. What is the solar mass of this black hole?

b. Express this solar mass in base 10.

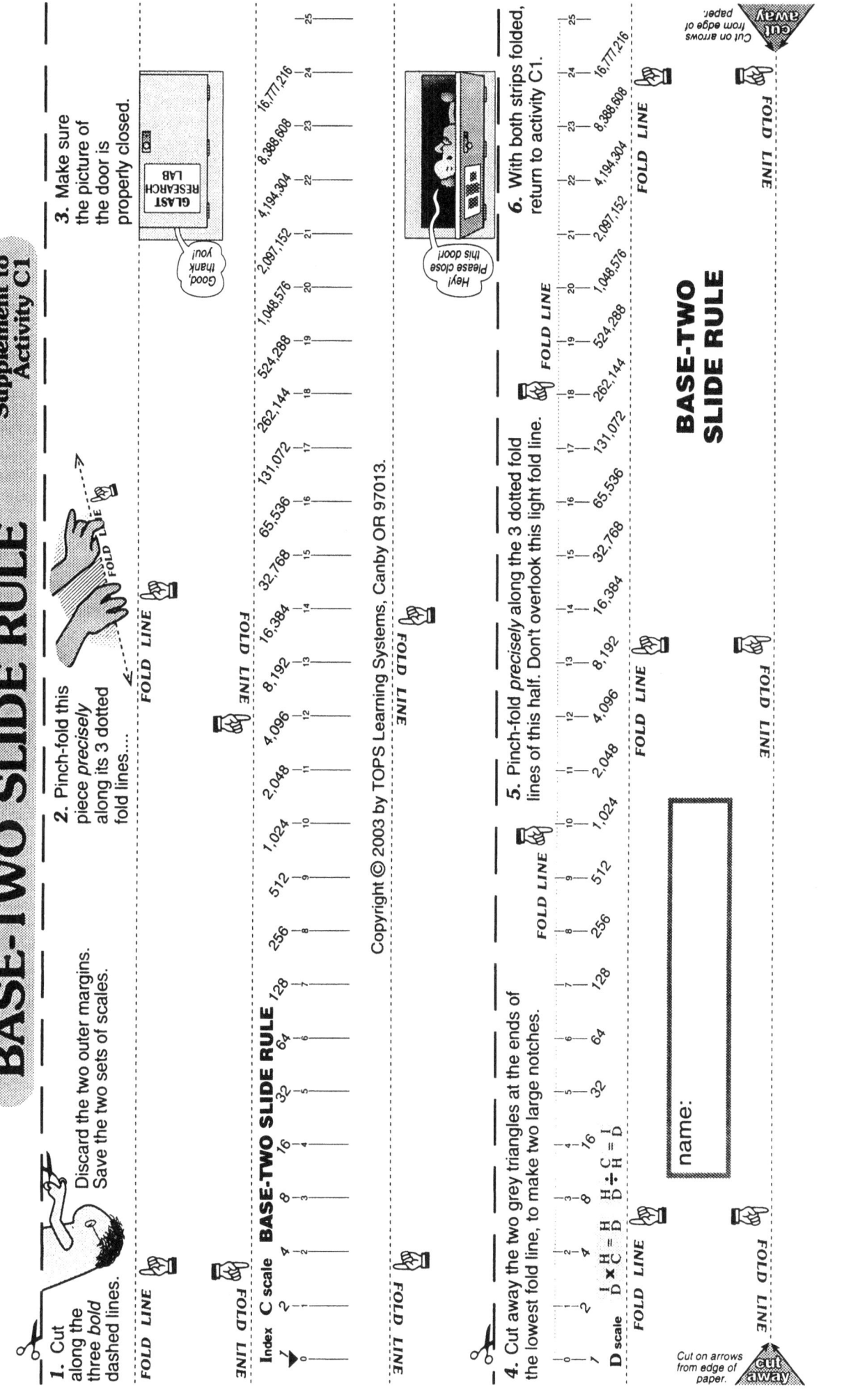

LOG RULER

Activity D1

1. Get the *Log Ruler* supplement. Fold as directed.
2. Fit it with a hairline, as before.

3. Practice reading its **log** scale (copied below), to 3 decimal places.

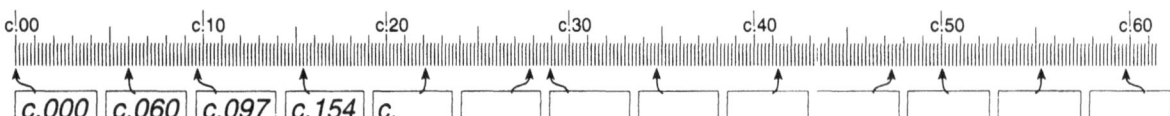

c.000 c.060 c.097 c.154 c.

a. Compare answers with a friend.
b. How many figures (not counting "c"), are **certain**?_____ **Uncertain**?_____

4. Practice reading its **antilog** scale (copied below) to 3 or 4 figures, as shown.

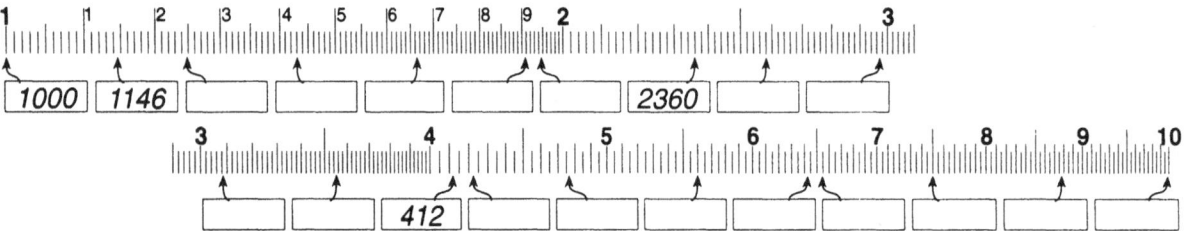

1000 1146 _ _ _ 2360 _ _

_ _ 412 _ _ _ _ _ _ _

a. Compare answers with a friend.
b. How many figures are **certain** below antilog 4? _____ Above antilog 4? _____

5. Get the *Playing Scales* page. Play both cooperative games with a friend, as directed.

6. Find these logs (exponents).

Number of decimal places from standard position matches characteristic (c).

7. Find these antilogs (numbers).

$15.0 = 10^{1.176}$	log 15.0 = 1.176	$10^{0.618} = 4.15$	antilog 0.618 = 4.15
$150 = 10^{2.176}$	log 150 = 2.176	$10^{2.618} = 415$	antilog 2.618 = 415
$1.50 = 10$		$10^{1.618} =$	
$.150 = 10$		$10^{\overline{1}.618} =$	
$16.0 = 10$		$10^{1.673} =$	
$.016 = 10$		$10^{2.673} =$	
$5.14 = 10$		$10^{0.871} =$	
$514. = 10$		$10^{\overline{2}.871} =$	
$5.61 = 10$		$10^{2.960} =$	
$.561 = 10$		$10^{\overline{1}.960} =$	

Find antilog, read log. *Find log, read antilog.*

GLASToids! 8. Search for facts about gamma rays.

a. What are gamma rays?	c. What absorbs gamma rays?
b. How heavy are they and how fast do they travel?	d. Do they "fly" straight?

OVERVIEW / OBJECTIVES

Students will construct *Log Rulers*, finely calibrated in base-10 exponents and numbers (logs and antilogs). They'll practice reading these scales as accurately as possible, listing all certain figures plus one uncertain figure.

TIME: 1 hour

INTRODUCTION

1. Photocopy the scales below. Fold along the dotted lines to display them one at a time in A-B-C order. Ask students to estimate, by secret ballot on scrap paper, the position of each arrow. Observe how smaller calibrations increase accuracy but don't eliminate uncertainty in the last digit.
(A: 96 or 97. B: 96.5 or 96.6. C: 96.52 or 96.53.)

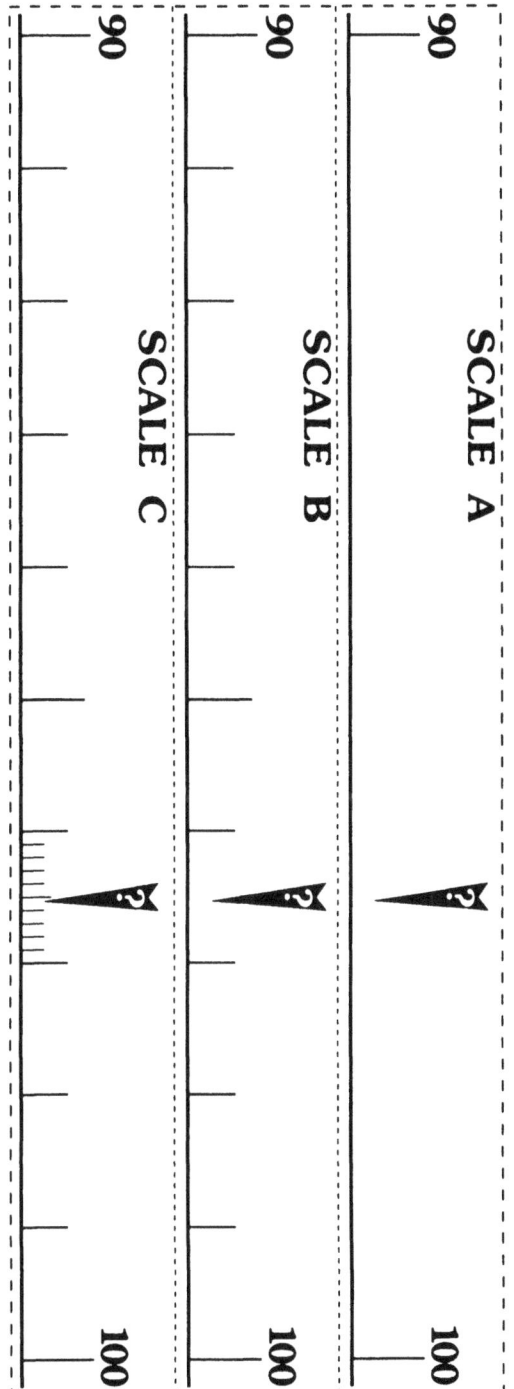

NOTES (this page)

5. Students will enjoy the cooperative game *Playing Scales* while learning much about measuring uncertainty. You might demonstrate the game as outlined here. You'll need 2 *Log Rulers*, 2 game pages, and a volunteer partner. Each player follows each step:

  (a) Circle a random log on the game page, then set this value with the hairline on the *Log Ruler*.

  (b) Trade papers and *Log Rulers*. Don't disturb hairlines!

  (c) Read the hairline, write the log, and compare it against the circled number.

  (d) If both measurements agree (within the limits of measuring uncertainty), blacken the small circle.

6-7. The log ruler supplies the mantissa, but *you* supply the characteristic (say where the decimal goes). See the Introduction on page 49.

ANSWERS (this page)

3. ... c.*221*, c.*278*, c.*289*, c.*347*, c.*413*, c.*474*, c.*500*, c.*552*, c.*598*

3b. ... *2 certain* figures, *1 uncertain* figure.

4. ... *1248, 1430, 1662, 1907, 1945,* 2360, *2580, 2970*

   ... *3085, 3550,* 412, *423, 476, 560, 643, 654, 750, 877, 998*

4b. ... *3 certain* below antilog 4; *2 certain* above.

5. Did students check both boxes at the bottom of Playing Scales? You might verify student skill levels by moving the hairline on a *Log Ruler* to any random position, then asking students to read both scales. Encourage "uncertain" students to play additional games.

6. $15.0 = 10^{1.176}$ $\log 15.0 = 1.176$
$150 = 10^{2.176}$ $\log 150 = 2.176$
$1.50 = 10^{0.176}$ **$\log 1.50 = 0.176$**
$.150 = 10^{\overline{1}.176}$ **$\log .150 = \overline{1}.176$**
$16.0 = 10^{1.204}$ **$\log 16.0 = 1.204$**
$.016 = 10^{\overline{2}.204}$ **$\log .016 = \overline{2}.204$**
$5.14 = 10^{0.711}$ **$\log 5.14 = 0.711$**
$514. = 10^{2.711}$ **$\log 514 = 2.711$**
$5.61 = 10^{0.749}$ **$\log 5.61 = 0.749$**
$.561 = 10^{\overline{1}.749}$ **$\log .561 = \overline{1}.749$**

7. $10^{0.618} = 4.15$ antilog $0.618 = 4.15$
$10^{2.618} = 415$ antilog $2.618 = 415$
$10^{1.618} = \mathbf{41.5}$ **antilog $1.618 = 41.5$**
$10^{\overline{1}.618} = \mathbf{.415}$ **antilog $\overline{1}.618 = .415$**
$10^{1.673} = \mathbf{47.1}$ **antilog $1.673 = 47.1$**
$10^{2.673} = \mathbf{471.}$ **antilog $2.673 = 471.$**
$10^{0.871} = \mathbf{7.43}$ **antilog $0.871 = 7.43$**
$10^{\overline{2}.871} = \mathbf{.0743}$ **antilog $\overline{2}.871 = .0743$**
$10^{2.960} = \mathbf{912.}$ **antilog $2.960 = 912.$**
$10^{\overline{1}.960} = \mathbf{.912}$ **antilog $\overline{1}.960 = .912$**

GLASTOIDS (this page)

8a. *Gamma rays are photons of electomagnetic energy, millions to billions of times more powerful than visible light.*

8b. *Gamma rays have no mass. They travel at the speed of light.*

8c. *Earth's atmosphere absorbs gamma rays, converting them to showers of electrons and positrons.* (GLAST also absorbs gamma rays!)

8d. *Yes, gamma rays are not deflected by magnetic or gravitational fields.*

NOTES: **Activity D1**

LOG RULER

Supplement to Activity D1

1. *Trim **both** sides.*

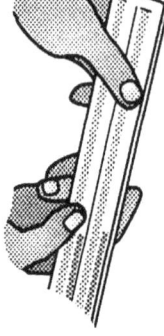

3. Fold over this **double layer** of paper by pinching precisely along this dashed line.

4. Fold over **all four layers** of paper by pinching precisely along this final dashed line.

Here's the second step…

2. Fold over this **single layer** of paper in half by pinching precisely along this line.

name:

LOG RULER

antilogs (ordinary numbers)

logs (base-ten exponents)

1. *Trim **both** sides.*

Copyright © 2003 by TOPS Learning Systems, Canby OR 97013.

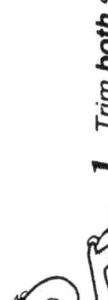

PLAYING SCALES

Supplement to Activity D1

PLEASE READ **ALL** INSTRUCTIONS **BEFORE** STARTING. (Review as needed while playing.)

Both you and your partner should do all steps simultaneously:

2. **(a)** Randomly circle one log listed below, left,
(b) Set the hairline on the **LOG** scale of your *Log Ruler*,
(c) Fold the top half of the paper to hide the list,
(d) Trade rulers and pages with your partner. Don't move the hairline!

1. Find a partner. Each fold your page on the vertical dashed line. Start with the **LOG** list face up.

LOG List

.632 ± 1	.000 ± 1	.458 ± 1	.675 ± 1
.343 ± 1	.621 ± 1	.666 ± 1	.489 ± 1
.554 ± 1	.432 ± 1	.388 ± 1	.738 ± 1
.010 ± 1	.743 ± 1	.584 ± 1	.842 ± 1
.421 ± 1	.824 ± 1	.999 ± 1	.921 ± 1
.165 ± 1	.965 ± 1	.109 ± 1	.128 ± 1
.776 ± 1	.176 ± 1	.751 ± 1	.244 ± 1
.887 ± 1	.288 ± 1	.863 ± 1	.055 ± 1
.598 ± 1	.098 ± 1	.537 ± 1	.317 ± 1
.909 ± 1	.309 ± 1	.953 ± 1	.291 ± 1
.219 ± 1	.023 ± 1	.211 ± 1	.405 ± 1

ANTILOG List

2084 ± 4	1055 ± 1	1765 ± 3	763 ± 1
537 ± 1	573 ± 1	521 ± 1	3775 ± 4
1478 ± 2	1365 ± 2	1458 ± 2	650 ± 1
888 ± 2	806 ± 2	850 ± 2	706 ± 1
1852 ± 3	1711 ± 3	1666 ± 3	3790 ± 4
439 ± 1	490 ± 1	466 ± 1	607 ± 1
2263 ± 4	3745 ± 5	1234 ± 2	755 ± 1
997 ± 2	968 ± 2	901 ± 2	3208 ± 5
2435 ± 4	2963 ± 4	2736 ± 4	634 ± 1
485 ± 1	450 ± 1	434 ± 1	1688 ± 3
3435 ± 5	2120 ± 4	3805 ± 4	543 ± 1

3. **(a)** Read the hairline setting on your partner's ruler, and write your answer below.
(b) Compare it to the number circled above. If your partner's setting and your answer **agree**, blacken the circle beside your written number below.
(c) To avoid confusion, cross out old circled numbers above. Then start a new round: circle a new number, set your hairline, and pass both to your partner.

4. Keep playing until both pages have at least **6 correct answers**. Then play the **ANTILOG** list on the right half of the page in the same way.

LOG Answers

ANTILOG Answers

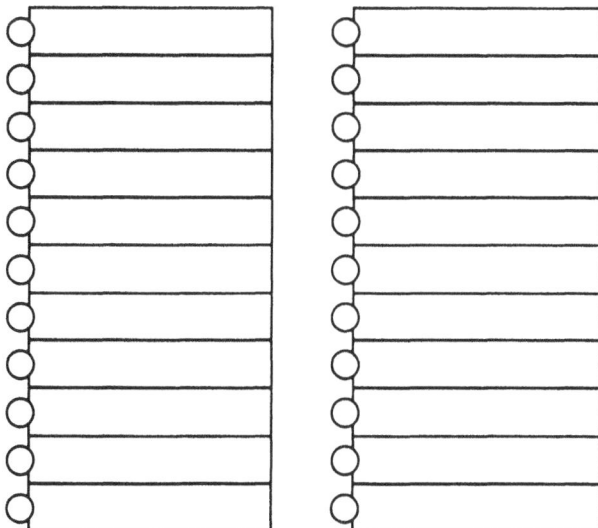

☐ We can read the **log scale** within acceptable limits of error.

☐ We can read the **antilog scale** within acceptable limits of error.

Copyright © 2003 by TOPS Learning Systems, Canby OR 97013

SCIENTIFIC NOTATION

1. Convert antilogs to logs. Follow these steps:

a. Start with an ordinary **N**umber (the **antilog**).
b. Convert to **scientific notation**.
c. Use **Log Ruler** to find **log** (exponent) to 3 decimal places.

2. Convert logs to antilogs. Follow these steps:

a. Start with the **log** (exponent).
b. Convert to **scientific notation**.
c. Use **Log Ruler** to find ordinary **N**umber (the **antilog**).

N (ANTILOG) →	scientific notation	**exp** (LOG)	**exp** (LOG)	scientific notation →	**N** (ANTILOG)
.0000282	2.82 × 10⁻⁵	$\bar{5}.450$	$\bar{5}.450$	2.82 × 10⁻⁵	.0000282
.00563	× 10⁻³		1.950	× 10¹	
3,980			1.800		
1.778			$\bar{4}.100$		
.251			2.200		
.01122			$\bar{6}.150$		
70,800			2.300		
355			0.900		
5,010,000			$\bar{2}.000$		
31,600			1.650		

LOG Answers! 6.700; $\bar{2}.050$; 2.550; 3.600; 0.250; $\bar{1}.400$; $\bar{3}.750$; 4.500; 4.850

ANTILOG Answers! 199.5 .0001259 44.7 89.1 158.5 .01 .000001413 63.1 7.94

3. For each boxed problem…
- ✓ Solve on a **Calc**ulator.
- ✓ State the log **Rule**.
- ✓ Do the **Log Work**.

(No Log Ruler needed! Just look up logs and antilogs in the above tables!)

Look up **LOGS** in this table... ...**ANTILOGS** on this side.

a. Calc: .251 × 355 = ☐

Rule: log A + log B = log AB

Log Work: .251 × 355 = ☐

$\bar{1}.400$ +   =

b. Calc: .00563 × 31,600 × .251 = ☐

Rule:

Log Work: .00563 × 31,600 × .251 =

OVERVIEW / OBJECTIVES

Students will convert antilogs to logs, and logs to antilogs, using scientific notation as an intermediate step. They will thereby develop a look-up table for solving math problems by using logarithms.

TIME: 1 hour and 10 minutes.

INTRODUCTION

A: Tape a *Log Ruler* up on your board. Label four exponents along the lower scale, as shown in *A*, below.

Examine log **c.6** below. This log contains a generalized **c** (the *characteristic*). By substituting some whole number for **c**, we get some decimal variation of 4. For example, if **c** = 0, the antilog is simply 4; if **c** = 1, the decimal place shifts one place right, making an antilog of 40; further, if **c** = 2, the antilog is 400; and so on.

B: Write *antilogs* directly above the ruler as various multiples of ten that correspond to different values of the characteristic **c**.

C: List corresponding logs (base-10 exponents) underneath, showing how **c** *defines the decimal place* of each respective number above it.

D: List equivalent scientific notation beneath the logs, to show how **c** equals the final exponent in this notation.

B:

100	200	400	800
10	20	40	80
1	2	4	8
.1	.2	.4	.8

antilogs (numbers) / logs (exponents)

A: c.0   c.3   c.6 (characteristic ↕ mantissa)   c.9

C:
$10^{2.0}$	$10^{2.3}$	$10^{2.6}$	$10^{2.9}$
$10^{1.0}$	$10^{1.3}$	$10^{1.6}$	$10^{1.9}$
$10^{0.0}$	$10^{0.3}$	$10^{0.6}$	$10^{0.9}$
$10^{\bar{1}.0}$	$10^{\bar{1}.3}$	$10^{\bar{1}.6}$	$10^{\bar{1}.9}$

D:
1×10^2	2×10^2	4×10^2	8×10^2
1×10^1	2×10^1	4×10^1	8×10^1
1×10^0	2×10^0	4×10^0	8×10^0
1×10^{-1}	2×10^{-1}	4×10^{-1}	8×10^{-1}

NOTES (this page)

1-2. An answer key is provide at the bottom of each table so students can check their work at each step.

3. These log problems are also self-checking, since log work must agree with calculator answers. We've rounded calculator answers to significant figures. Students typically report all digits on their display.

ANSWERS (this page)

1.
.00563	5.63×10^{-3}	$\bar{3}.750$
3,980	3.98×10^3	3.600
1.778	1.778×10^0	0.250
.251	2.51×10^{-1}	$\bar{1}.400$
.01122	1.122×10^{-2}	$\bar{2}.050$
70,800	7.08×10^4	4.850
355	3.55×10^2	2.550
5,010,000	5.01×10^6	6.700
31,600	3.16×10^4	4.500

2.
1.950	8.91×10^1	89.1
1.800	6.31×10^1	63.1
$\bar{4}.100$	1.259×10^{-4}	.0001259
2.200	1.585×10^2	158.5
$\bar{6}.150$	1.413×10^{-6}	.000001413
2.300	1.995×10^2	199.5
0.900	7.94×10^0	7.94
$\bar{2}.000$	1.000×10^{-2}	.01
1.650	4.47×10^1	44.7

3a. Calc: **89.1**
Rule: given
Log Work: **89.1**
$\bar{1}.400 + 2.500 = 1.950$

b. Calc: **44.6**
Rule: **log A + log B + log C = log ABC**
Log Work: **44.7**
$\bar{3}.750 + 4.500 + \bar{1}.400 = 1.650$

ANSWERS (next page)

3c. Calc: **158.5**
Rule: **log A − log B = log A/B**
Log Work: **158.5**
$6.700 - 4.500 = 2.200$

d. Calc: **.0100**
Rule: **log A + log B = log AB**
Log Work: **.0100**
$0.250 + \bar{3}.750 = \bar{2}.00$

e. Calc: **199.7**
Rule: **log A − log B = log A/B**
Log Work: **199.5**
$2.550 - 0.250 = 2.300$

f. Calc: **63.1**
Rule: given
Log Work: **63.1**
$3.600/2 = 1.800$

g. Calc: **7.94**
Rule: **log $^n\sqrt{A}$ = log A/n**
Log Work: **7.94**
$3.600/4 = .900$

h. Calc: **.0001259**
Rule: given
Log Work: **.0001259**
$2 \times \bar{2}.050 = \bar{4}.100$

i. Calc: **.000001412**
Rule: **log A^n = n log A**
Log Work: **.000001413**
$3 \times \bar{2}.050 = \bar{6}.150$

j. Calc: **199.5**
Rule: given
Log Work: **199.5**
$4.850/2 - 0.250/2 = 2.300$

4. ...addition, ...subtraction, ...multiplication, ...division

GLASTOIDS (next page)

4. Calculator solution:
$E = mc^2$
$E = (1.67 \times 10^{-27} kg)(3 \times 10^8 m/s)^2$
$= 15.03 \times 10^{-11}$ kg m²/sec²
$= 1.503 \times 10^{-10}$ joules

Log Solution:
$E = mc^2$
$\log E = \log mc^2$
$= \log m + 2 \log c$
$= \bar{27}.222 + 2(8.477)$
$= \bar{27}.222 + 16.954$
$= \bar{10}.176$

antilog log E = antilog 10.176
$E = 1.5 \times 10^{-10}$ joules

NOTES: **Activity D2**

SCIENTIFIC NOTATION

Activity D2

c. Calc: $5,010,000 \div 31,600 =$ ____

Rule: $\log A - \log B = \log A/B$

Log Work:
$5,010,000 \div 31,600 =$ ____

log N → log N → antilog exp ↑

d. Calc: $1.778 \times .00563 =$ ____

Rule: ____

Log Work:
$1.778 \times .00563 =$ ____

log N → log N → antilog exp ↑

e. Calc: $355 \div 1.778 =$ ____

Rule: ____

Log Work:
$355 \div 1.778 =$ ____

log N → log N → antilog exp ↑

f. Calc: $\sqrt[2]{3,980} =$ ____

Rule: $\dfrac{\log A}{n} = \log \sqrt[n]{A}$

Log Work:
$3,980^{1/2} =$ ____

log N → antilog exp ↑

g. Calc: $\sqrt{3,980} =$ ____

Rule: ____

Log Work:
$3,980^{1/4} =$ ____

log N → antilog exp ↑

h. Calc: $.01122^2 =$ ____

Rule: $n \log A = \log A^n$

Log Work:
$.01122^2 =$ ____

log N → antilog exp ↑

i. Calc: $.01122^3 =$ ____

Rule: ____

Log Work:
$.01122^3 =$ ____

log N → antilog exp ↑

j. Calc: $(70,800/1.778)^{1/2} =$ ____

Rule: $\dfrac{1}{n}\log A - \dfrac{1}{n}\log B = \log (A/B)^{1/n}$

Log Work:
$(70,800/1.778)^{1/2} =$ ____

log N → log N → antilog exp ↑

4. Logs reduce the complexity of a calculation "one notch." This means:

Multiplication *becomes* _____

Division *becomes* _____

Raising to a Power *becomes* _____

Taking a Root *becomes* _____

5. *GLASToids!* As an atom of hydrogen spins into a black hole at near light speed (3×10^8 meters per second), its entire mass (1.67×10^{-27} kilograms) is converted to energy. Apply Einstein's famous equation, found in these *GLASToids*, to work out the energy released (in joules). Show your work on another paper or the back of this one. Extra credit if you also use logs!

CLASSIC LOG SCALES

CLASSIC SLIDE RULE

Supplement to Activity D3

1. Cut along the three *bold* dashed lines.

Discard the two outer margins. Save the two sets of scales.

2. Pinch-fold *precisely* on the six dotted fold lines....

3. Make sure the picture of the door is properly closed.

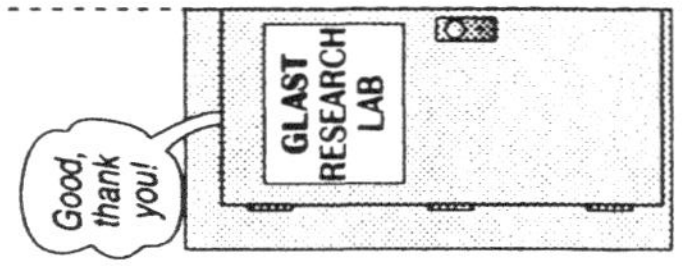

Good, thank you!

GLAST RESEARCH LAB

4. Cut away the two grey triangles at the ends of the lowest fold line, to make two large notches.

5. Pinch-fold *precisely* along the 3 dotted fold lines of this half. Don't overlook this light fold line.

$$I \times H = H \quad H + C = I$$
$$D \times C = D \quad D + H = D$$

6. With both strips folded, return to activity D3.

Hey! Please close this door!

C SCALE

D SCALE

CLASSIC SLIDE RULE

CLASSIC SLIDE RULE

LEFT INDEX
RIGHT INDEX

FOLD LINE

Copyright © 2003 by TOPS Learning Systems, Canby OR 97013.

name:

Cut on arrows from edge of paper.

cut away

PAGE 51

CLASSIC SLIDE RULE

Activity D3
1 of 2 pages

1. Get the *Classic Slide Rule* supplement. Fold as directed.

2. Fit it with a hairline as usual.

3. Practice these simple multiplications...

...using the **LEFT index**.	...using the **RIGHT index**.	...using **EITHER index**.
2 x 2 = ____	9 x 9 = ____	8 x 2 = ____
3 x 3 = ____	8 x 8 = ____	4 x 2 = ____
2 x 3 = ____	7 x 7 = ____	5 x 2 = ____
2 x 4 = ____	6 x 6 = ____	6 x 2 = ____

a. When using a slide rule, how do you know which index to use?

b. Which problem can be calculated using **both** indexes?

4. Keep **multiplying** by 2, writing each answer in its box. Use significant figures: what can actually read + 1 estimated figure.

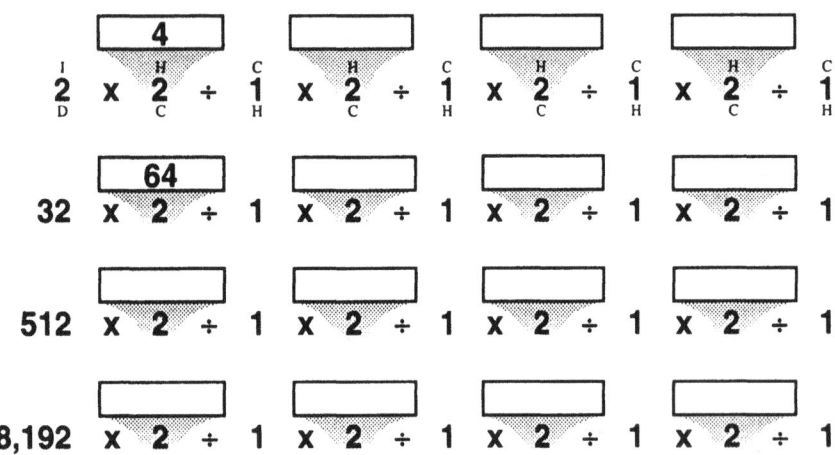

Move hairline to keep multiplying!

A *Classic Slide Rule* tells you much about the answer but not everything. Explain.

5. Keep **dividing** by 2, writing each answer in its box. Use significant figures: what can actually read + 1 estimated figure.

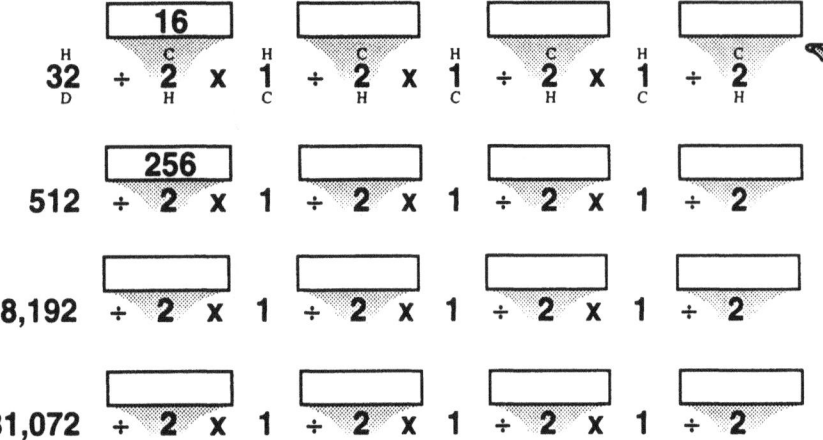

Move C-scale to keep dividing!

Can you ever exceed the capacity of a *Classic Slide Rule*? Explain.

6. **Above** each box, write the answer given on your *Base-Two Slide Rule.*

Compare the numbers above each box to the numbers inside. How accurate is your *Classic Slide Rule*?

OVERVIEW / OBJECTIVES

Students will construct *Classic Slide Rules* like those in common use until the mid 1970's. They'll use them like calculators, learning much about reading scales, significant figures and estimating decimal places.

TIME: 1 hour and 30 minutes

INTRODUCTION

1. Read the major divisions on a meter stick to your class in tenths of meters: .000 m, .100 m, .200 m,.... Now, instead of using meters, let's define these same numbers as base-10 exponents. Our meter stick then becomes a finely divided exponent scale, with the major divisions being: $10^{.000}$, $10^{.100}$, $10^{.200}$, ...

Ask students to use masking tape and a marking pen to place the integers 2 through 9 precisely on their exponent values, thereby developing a logarithmic scale. The integer 2, for example, lands on .301 m; 3 belongs at .477 m, and so on, up to 9 at .954 m. (Find these equivalents on the *Log Ruler* or *Log Tape*.) Finally, wrap tape over each end, and label with 1 on each outside edge.

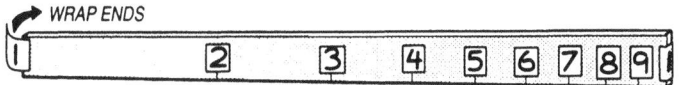

Calibrate a second meter stick like the first. Use rolls of masking tape to fix one scale to your board or wall, then rest the second above it.

Using thumb and finger as a "hairline," demonstrate how *both* the left and right index now multiply these numbers (or any other multiple of 10).

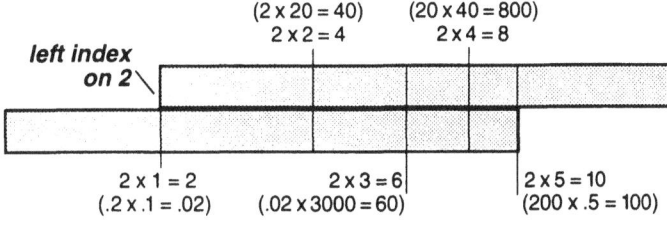

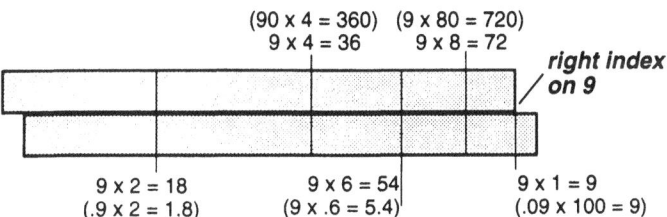

Draw a heavy "hairline" at 60 on the D-scale. Slide the C-scale to divide 60 by 6, then 5, then 4, then 3, then 2. Read each result on the right index.

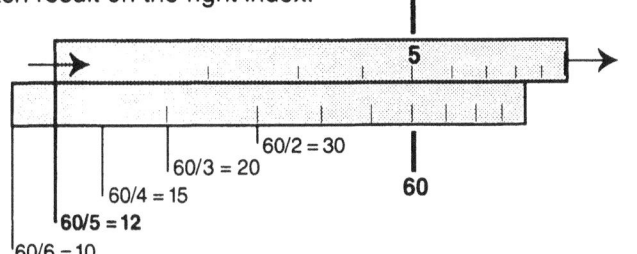

NOTES (this page)

2. "Two times two" equals "four" on a *Classic Slide Rule*. This is what students will see unless they do this operation *very* carefully, and look *very* closely. Then they might actually read 2.01 or 2.02. This error is inherent within the instrument, unavoidably introduced in the process of printing and duplicating the scale. The slide rule still works beautifully within a margin of error of about 1%. *All* scientific instruments carry some margin of error.

4-5. This practice allows students to experience how slide rules alternate between hairline multiplication and C-scale division. Knowing what answers should be (from their *Base-2 Slide Rules*), students can easily detect and self-correct operational errors. Beyond 3 digits (after 512), students learn to add zeros to hold the decimal place.

ANSWERS (this page)

3. | LEFT index: | RIGHT index: | EITHER index: |
    |---|---|---|
    | 2 x 2 = **4** | 9 x 9 = **81** | 8 x 2 = **16** |
    | 3 x 3 = **9** | 8 x 8 = **64** | 4 x 2 = **8** |
    | 2 x 3 = **6** | 7 x 7 = **49** | 5 x 2 = **10** |
    | 2 x 4 = **8** | 6 x 6 = **36** | 6 x 2 = **12** |

3a. *Use whichever index keeps both multiplied numbers on the D scale, with no number hanging off an end. (If one index doesn't work, the other index surely will.)*

3b. *5 x 2 = 10. (Numbers like 1, 10, 100, .1, .01, .001, etc, are the only ones to appear twice, on both ends of the D scale.)*

4. (Variations in the 3rd and 4th figure are OK!)

4	8	16	32
64	*128*	*256*	*513*
1021	*2040*	*4100*	*8200*
16400	*32800*	*65600*	*131000*

A classic slide rule only gives the numbers, but doesn't say where to place the decimal.

5.

16	*8*	*4*	*2*
255	*127*	*64*	*32*
4060	*2040*	*1022*	*512*
65800	*32800*	*16350*	*8200*

Since the slide rule does not define the decimal place, its capacity cannot be exceeded.

6. (Students should copy exact answers directly off their *Base-2 Slide Rules* and write them *above* each box):

4	8	16	32
64	128	256	512
1024	2048	4096	8192
16384	32768	65536	131072

16	8	4	2
256	128	64	32
4096	2048	1024	512
65536	32768	16384	8192

The numbers *above* the boxes are exact. The numbers *inside* the boxes are accurate to 3 significant figures, with measuring uncertainty in the last digit.

NOTES: **Activity D3**

CLASSIC SLIDE RULE

Activity D3

7. Complete these problems on your *Classic Slide Rule*.

PROBLEM	PLACE DECIMAL (BY ROUGH CALCULATION)	SLIDE RULE ANSWER	CALCULATOR CHECK
$9.350 \times 49.50 = ?$	$10 \times 50 = 500$	463	462.825
$19.75 \times 425 = ?$			
$12.00 \times 8.00 \times 19.00 = ?$			
$\dfrac{62,200.}{29.20} = ?$			
$\dfrac{2,500.}{2.000 \times 3.000 \times 4.000} = ?$			
$(2.95)^3 = ?$			
$\dfrac{83.2 \times 77.3}{215} = ?$			

8. Another way to track decimals is...
(USING SCIENTIFIC NOTATION)

$1230. \times .00321 = ?$	(mult. N's) $(1.23 \times 10^3)(3.21 \times 10^{-3}) \approx 4 \times 10^0 = 4$ (add exp's)		
$1975 \times 314 = ?$			
$(202)^2 = ?$			
$(.292)^3 = ?$			

9. Your *Classic Slide Rule* adds and subtracts logs to multiply and divide numbers, yet no logs are printed on it anywhere! Where are they hiding?

10. Notice how the bold numbers (1, 2, 3, ...9) are arranged on your *Classic Slide Rule*.

    a. Compare this pattern to the antilog scale on your *Log Tape*.

    b. Compare this pattern to the smallest calibrations on your *Multiplying Slide Rule*.

GLASToids!

11. A large black hole resides at the center of our own Milky Way Galaxy.
    a. Search the GLASToids to find its solar mass and the mass of our sun.

    b. Use your slide rule to calculate its gram mass. Show your work.

NOTES (this page)

7-8. A *Classic Slide Rule* scale can be accurately read to 4 significant figures on the left side of the instrument (where the scale is expanded), otherwise to 3 significant figures. When actually multiplying or dividing numbers, however, inherent error in the slide rule scale introduces an error of about 1%. Thus, slide rule answers and calculator answers will differ in the third figure, even though students may sometimes legitimately report answers with 4 figure accuracy.

Students who have little or no experience with measuring uncertainty and significant figures generally report all the figures they can read on their calculator display, whether the numbers are significant or not. In the answer key below, we have rounded off calculator answers to the same number of significant figures as stated in the problem, then supplied "all the numbers" to the right.

10b. If you directed your class to skip STRAND A lessons, ask students to skip this question as well. You may wish to show them the small logarithmic patterns on the *Multiplying Slide Rule* master on page 23. These are in direct proportion to the larger patterns on the classic slide rule.

ANSWERS (this page)

7, 8. See answers listed below.

9. *The logs are hiding "under" the numbers. The position that each number occupies on the logarithmic scale is an expression of the magnitude of its base-10 exponent or log. (These exponents are not hidden on the Log Ruler.)*

10a. *The bold numbers on the Classic Slide Rule occupy the same relative positions as the bold numbers printed on the antilog scale of the Log Tape between 1 and 10.*

10b. *The bold numbers on the Classic Slide Rule occupy the same relative positions as the tiny numbers printed on the Base-10 Slide Rule. The scales are different, but the proportions are the same.*

7.

PROBLEM	PLACE DECIMAL	SLIDE RULE ANSWER	CALCULATOR CHECK
$9.350 \times 49.50 = ?$	$10 \times 50 = 500$	463	462.8 (462.825)
$19.75 \times 425 = ?$	$20 \times 400 = 8{,}000$	8420	8390 (8393.75)
$12.00 \times 8.00 \times 19.00 = ?$	$10 \times 10 \times 20 = 2{,}000$	1815	1824 (1824)
$\dfrac{62{,}200.}{29.20} = ?$	$\dfrac{60{,}000}{30} = 2{,}000$	2138	2140 (2130.137)
$\dfrac{2500.}{2.000 \times 3.000 \times 4.000} = ?$	$\dfrac{2{,}500}{25} = 100$	103.4	104.2 (104.16667)
$(2.95)^3 = ?$	$3^3 = 27$	25.7	25.7 (25.672375)
$\dfrac{83.2 \times 77.3}{215} = ?$	$\dfrac{80 \times 80}{200} = \dfrac{64}{2} = 32$	29.7	29.9 (29.913302)

8.

PROBLEM	PLACE DECIMAL	SLIDE RULE ANSWER	CALCULATOR CHECK
$1230. \times .00321 = ?$	$(1.23 \times 10^3)(3.21 \times 10^{-3}) \approx 4 \times 10^0 = 4$	3.96	3.95 (3.9483)
$1975 \times 314 = ?$	$(1.975 \times 10^3)(3.14 \times 10^2) \approx 6 \times 10^5 = 600{,}000$	622,000	620,000 (620,150)
$(202)^2 = ?$	$(2.02 \times 10^2)(2.02 \times 10^2) \approx 4 \times 10^4 = 40{,}000$	40,800	40,800 (40,804)
$(.292)^3 = ?$	$(2.92 \times 10^{-1})(2.92 \times 10^{-1})(2.92 \times 10^{-1}) \approx 27 \times 10^{-3} = .027$	.0250	.0249 (.0248971)

GLASTOIDS

11a. Solar mass of black hole = **2.6 million suns**
Mass of sun = **1.99×10^{33} grams**

11b. **$(2.6 \times 10^6 \text{ suns})(1.99 \times 10^{33} \text{ grams/sun}) = 5.2 \times 10^{39}$ grams**

NOTES: **Activity D3**

SLIDE RULE GRAPH

Activity E1

1. Get the *Slide Rule Graph* supplement.

2. Notice how both plotted points correspond to numbers and logs on your **Log Tape**. Plot more ordered pairs (**N**, **log N**), for each **Number** printed on the horizontal axis.

Find all log N's on your log tape.

3. Circle each point. Draw a smooth curved line to connect the circles, but don't draw through your plotted points.

4. Extend your graph line below the horizontal axis to show its downward direction. As **log N** becomes "very" negative, what happens to **N**?

5. Extend parallel lines from each circled data point, across the white "alley", to the left axis. Be accurate.

6. Examine the line patterns you've drawn. Where have you seen them before?

7. Label each log position (pencil line) with its corresponding **Number** (antilog). Write directly on your pencil lines.

8. Cut the white alley along the dashed line in the center. Fold each section along the "arrow" lines... ...Tape the *D scale* (narrow strip) to your table, with your written numbers facing up.

9. Bring the folded edges together to make a hand-drawn slide rule! How does it multiply and divide?

10. Calibrate the **D scale** between 10 and 100 with 23 more numbers! *(Hint: Since 6 x 2 = 12, you can mark and label 12. Keep multiplying and calibrating: 14, 16, 18, ... Use neat, small numbers.)*

GLASToids!

11. GLAST must do 4 things to "see" gamma rays. Name a GLAST system that does each task:

(1) Determine if the particle is really a gamma ray, and not a much more common cosmic ray.

(2) Determine what direction it comes from.

(3) Measure its energy.

(4) Communicate all this information to Earth.

GLAST is NOT a simple telescope with mirrors and lenses.

PAGE 56 LOG GRAPHS   Copyright © 2003 by TOPS Learning Systems, Canby OR 97013

OVERVIEW / OBJECTIVES

Using their *Log Tapes* as a reference for ordered pairs, students will graph positive numbers as a function of their base-10 logarithms. They'll extend each plotted point to the vertical axis, thereby generating a logarithmic scale that cuts and folds into an improvised slide rule.

TIME: 40 minutes

INTRODUCTION

Graph how log N becomes negative as N drops below 1. Help students understand the simple relationship between "bar logs" (with positive mantissas), and completely negative exponents. (Students won't plot negative logs until the next activity, but will be asked to describe trends below N = 1.)

Develop a data table and graph with logs rounded off to one or two figures as shown. Begin by completing only the left side of the table. Ask your class to consult their *Log Tapes* to fill in the right side, then plot the ordered pairs.

As you develop the graph line, emphasize these techniques: **(a)** Circle each data point. **(b)** Connect circles with a smooth curve. **(c)** Keep the graph line outside the circles. Don't obscure the data points.

ANSWERS (this page)

4. *As log N gets more and more negative, N approaches 0.* (Check to see that students have extended their graph lines below zero, as shown.)

6. *These line patterns are logarithmic, showing up on the Base-10 Slide Rule, the Log Tape, the Log Ruler, and the Classic Slide Rule.*

9. *This slide rule multiplies and divides numbers by adding and subtracting their logs (base-10 exponents).*

10. Students should calibrate the D-scale with these additional numbers: *12, 14, 15, 16, 18, 21, 24, 25, 27, 28, 32, 35, 36, 42 45, 48, 49, 54, 56, 63, 64, 72, 81.*

GLASToids
(1) *Anticoincidence System*   (3) *Calorimeter*
(2) *Precision Tracker*   (4) *On-Board Computers and Antennae*

MATERIALS (additional)
☐ A ruler or straightedge.

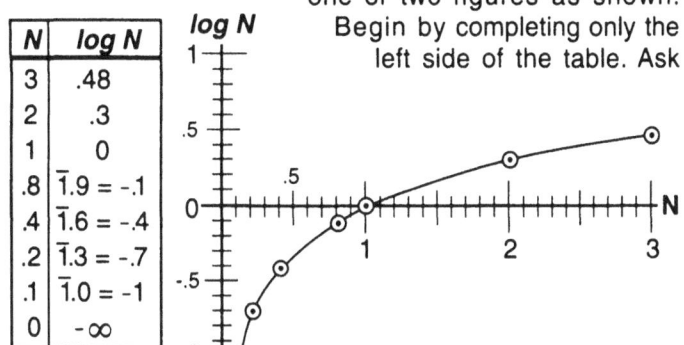

N	log N
3	.48
2	.3
1	0
.8	$\bar{1}.9 = -.1$
.4	$\bar{1}.6 = -.4$
.2	$\bar{1}.3 = -.7$
.1	$\bar{1}.0 = -1$
0	$-\infty$

Slide Rule Graph:

NOTES: **Activity E1**

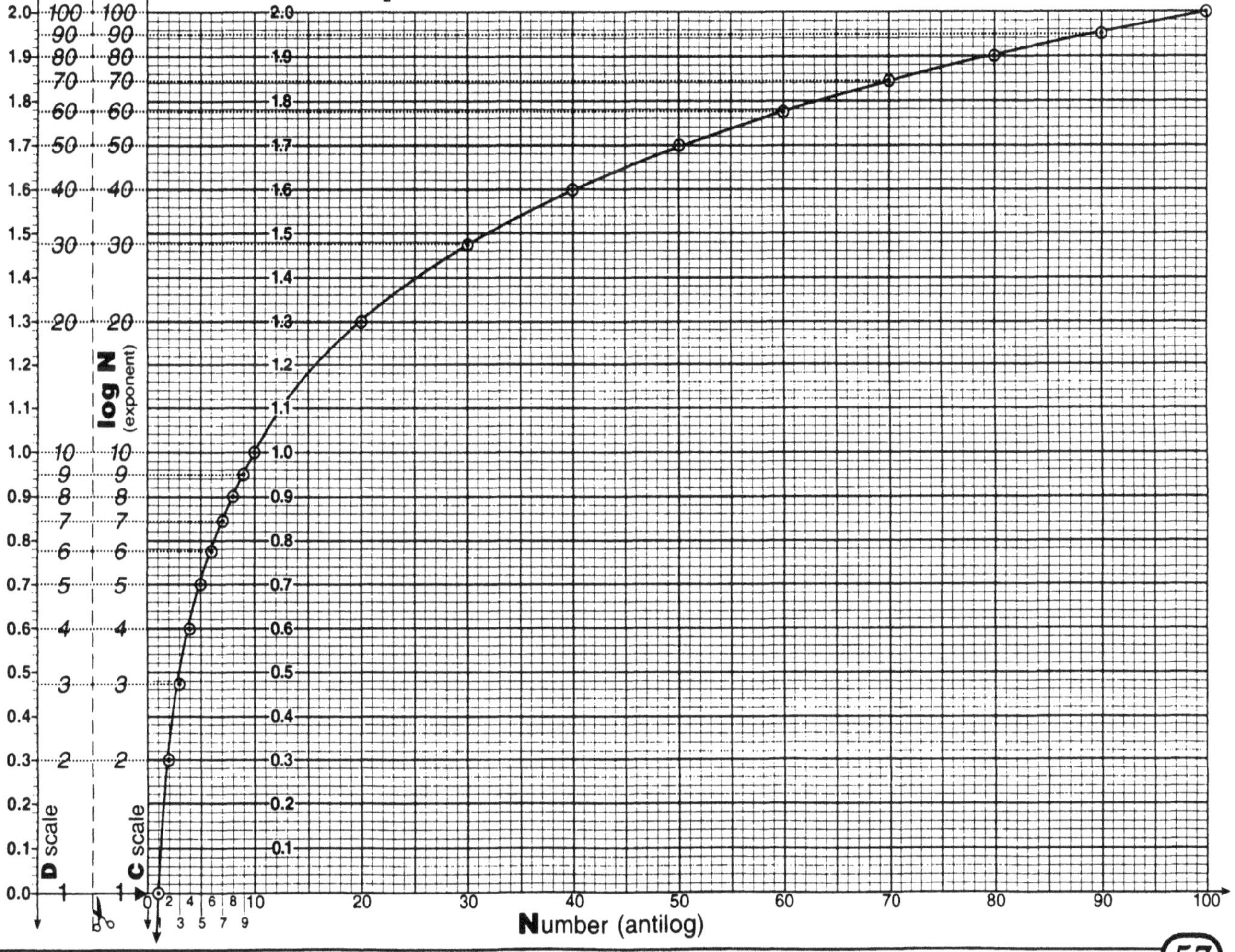

INVERSE FUNCTIONS

Activity E2
1 of 1 page

1. Get the supplement labeled *Inverse Functions*.

2. Fill in "y" values for the function **y = x**. (This is the top line of the data table.)

Function	Ordered Pairs
y = x	(-2, -2) (0, 0) (2, 2) (5,)

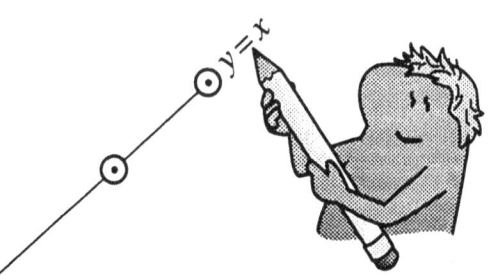

3. Plot and circle each ordered pair on the grid below this data table. Connect the circles with a smooth straight line, but don't draw through your data points.

4. Label this graph line **y = x**.

5. Repeat steps 2, 3, and 4 in a similar manner for all remaining functions. Use your *Base-two Slide Rule* and *Log Tape* as a reference.

y = x						
y = 2^x	(3,8)	(2,)	(1,)	(0,)	(-1,)	(-2,)
y = 10^x	(1,)	(.8,)	(.60,)		(0,)	(-0.155,)
y = $\log_2 x$	(8,)	(4,)		(1,)		
y = log x	(10,)	(7,)			(1,)	(.7,)

6. Inverse functions, when graphed, look like mirror-image reflections.

a. Which are inverse pairs? About what axis do these equations reflect?

b. You can identify inverse functions by comparing their ordered pairs. Explain.

7. Search the GLASToids for 8 astronomical phenomena that GLAST will study.

OVERVIEW / OBJECTIVES

Using their *Log Tapes* and *Base-2 Slide Rules* as references, students will graph exponential functions and log functions in base-10 and base-2. They'll discover that exponential and log functions are inverse, reflecting across the y = x axis as mirror images.

TIME: 50 minutes

INTRODUCTION

On your board or overhead, develop data tables and sketch graph lines for this pair of **inverse functions**.

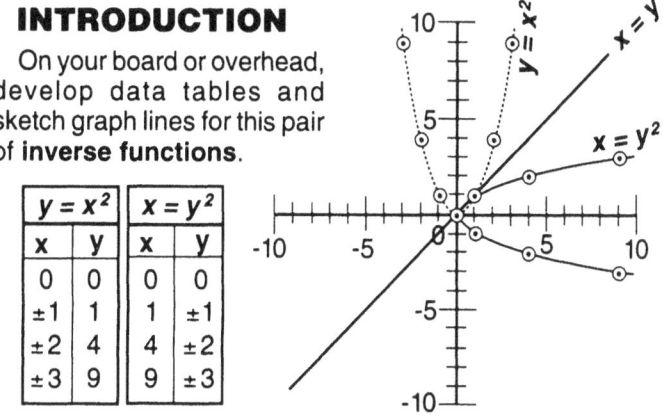

$y = x^2$		$x = y^2$	
x	y	x	y
0	0	0	0
±1	1	1	±1
±2	4	4	±2
±3	9	9	±3

How can you identify inverse functions? They interchange their (x,y) variables, and graph as mirror image reflections along the x = y axis.

ANSWERS (this page)

6a. These functions are inverse:
  $y = 2^x$ and $y = \log_2 x$
  $y = 10^x$ and $y = \log x$
  Both reflect across the y = x axis.

6b. Ordered pairs of inverse functions have their (x, y) values in reverse order.

GLASToids

Students should list these gamma ray sources in any order:

Solar Flares	Exotic Particles
Black Holes	Accretion Disks
Dark Matter	Active Galactic Nucleus (AGN)
Quasars	Gamma Ray Burst (GRB)

2-5. Inverse Functions

Function	Ordered Pairs (x,y)
$y = x$	(-2,-2) (0, 0) (2, 2) (5, 5) (10, 10)
$y = 2^x$	(3, 8) (2, 4) (1, 2) (0, 1) (-1, 1/2) (-2, 1/4)
$y = 10^x$	(1, 10) (.845, 7) (.602, 4) (.301, 2) (0, 1) (-0.155, .7) (-0.398, .4) (-0.699, .2) (-1, .1) (-2, .01)
$y = \log_2 x$	(8, 3) (4, 2) (2, 1) (1, 0) (1/2, -1) (1/4, -2)
$y = \log x$	(10, 1) (7, .845) (4, .602) (2, .301) (1, 0) (.7, -0.155) (.4, -0.398) (.2, -0.699) (.1, -1) (.01, -2)

NOTES: **Activity E2**

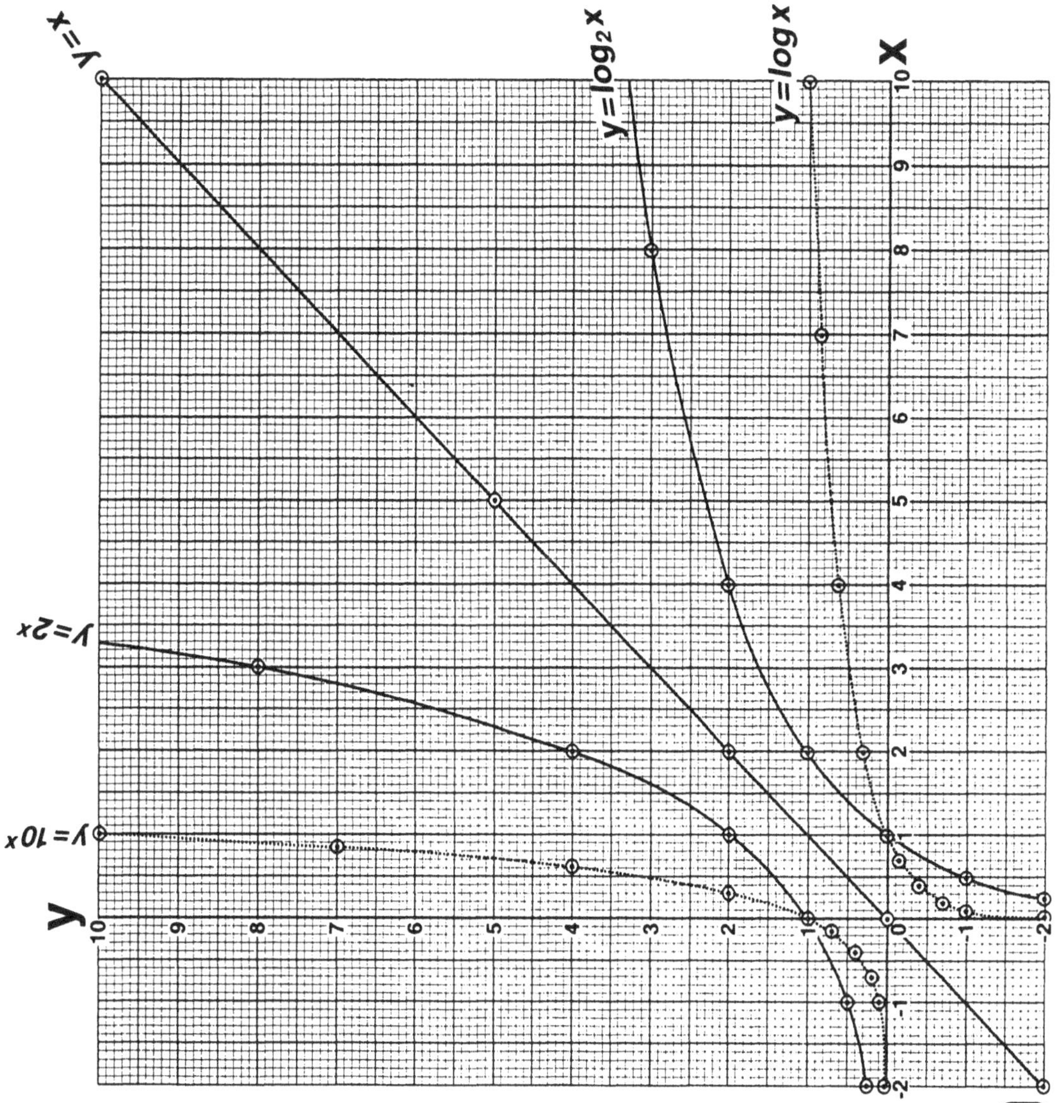

IRONING OUT CURVES

Activity E3
1 of 1 pages

1. Get the *Linear/Log Grids* supplement.

2. Finish numbering the **x** and **y** axes for both grids.

3. Notice the 1, 2, 10, 100, and 1000 on the y-axis of the *Log-Log Grid* are also labeled with their corresponding base-ten exponents.

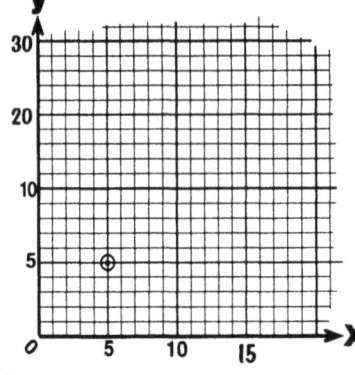
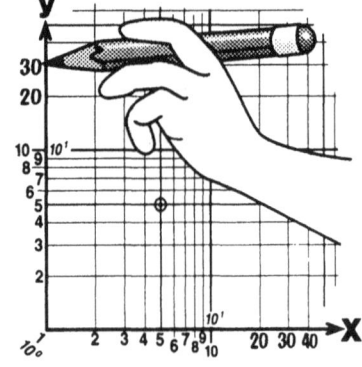

   a. Label 4, 8, 20, 40, 80, 200, 400, and 800 *(on the y axis only)*, in a similar way.

   b. On the *Log-Log Grid*, the _____ are evenly spaced, causing the corresponding _____ to be logarithmically spaced.

   c. Lines on the *Linear Grid* are numbered so the magnitude of each number is proportional to its distance from the origin (0,0). Explain, in a similar manner, how are lines on the *Log-Log Grid* numbered?

4. Fill in the top line of the data table for **y = x** as before, then plot your results on **both** grids. Circle your data points and label your graph lines in the margins.

5. Graph and label all remaining functions (on both graphs), in a similar manner.

6. Which functions are inverse pairs? What is the axis of reflection?

7. Each large square on the *Log-Log* graph spans one base-10 exponent (1 log cycle). Thus we can define the slope (m) of each graph line as a change in log values:

$$m = \Delta \log y / \Delta \log x.$$

Completed this table of log slopes for each equation:

$y = x$:	$m = 1/1 = 1$
$y = x^2$:	
$y = x^3$:	
$y = x^{1/2}$:	
$y = x^{1/3}$:	

b. What is the relationship between the slope of an equation on a *Log-Log Grid* and its power?

8. GLAST astronomers, more often than not, work with log-log grids rather than linear grids. Give two reasons why.

GLASToids!

OVERVIEW / OBJECTIVES

Students will graph second and third order functions, discovering an inverse relationship between squares and square roots; between cubes and cube roots. They'll graph these functions on both a linear grid (evenly spaced numbers), and a log-log grid (evenly spaced exponents). Graph lines that curve on linear grids transform into straight lines on the log-log grids, with slopes equal to their exponential powers.

TIME: 50 minutes

INTRODUCTION

Here's a a log-log grid, so named because the numbers (1, 10, 100, 1000, 10000), are spaced by the value of their logs on both axes. Notice how both equations graph as straight lines, where that the slope of the each line equals the value of the exponent:

$$m = \Delta \log y / \Delta \log x$$

$y = x$		$y = x^2$	
x	y	x	y
10^0	10^0	10^0	10^0
10^1	10^1	10^1	10^2
10^2	10^4	10^2	10^4
10^3	10^3		
10^4	10^4		

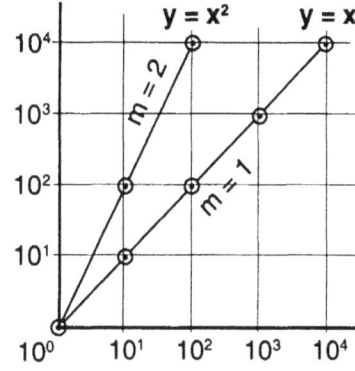

ANSWERS (this page)

2, 3a., 4, 5. See graphs and tables below.

3b. On a log-log grid, the <u>exponents</u> are evenly spaced, causing the corresponding <u>numbers</u> to be logarithmically spaced.

3c. Lines on the log-log grid are numbered so the magnitude of each number's exponent (log) is proportional to its distance from the origin (0,0).

6. These functions are inverse:
$y = x^2$ and $y = x^{1/2}$;
$y = x^3$ and $y = x^{1/3}$
They both reflect across the $y = x$ axis.

7a.
$y = x$: $m = 1/1 = 1$
$y = x^2$: $m = 2/1 = 2$
$y = x^3$: $m = 3/1 = 3$
$y = x^{1/2}$: $m = 1/2$
$y = x^{1/3}$: $m = 1/3$

7b. <u>The slope</u> ($m = \Delta \log y / \Delta \log x$) of each equation on a log-log grid <u>equals power to which it is raised.</u>

GLASToids

GLAST astronomers prefer log-log grids ...,
(1) when graphing huge ranges of data that wouldn't otherwise fit on a linear graph.
(2) to transform exponential curves into simple straight lines.

2, 3a., 4, 5.

Functions	Ordered Pairs (x,y)
$y = x$	(0, 0) (1, 1) (5, 5) (30, 30) (1000, 1000)
$y = x^2$	(0, 0) (1, 1) (2, 4) (3, 9) (4, 16) (5, 25) (10, 100) (30, 900)
$y = x^3$	(0, 0) (1, 1) (2, 8) (3, 27) (10, 1000)
$y = x^{1/2}$	(0, 0) (1, 1) (4, 2) (9, 3) (16, 4) (25, 5) (100, 10) (900, 30)
$y = x^{1/3}$	(0, 0) (1, 1) (8, 2) (27, 3) (1000, 10)

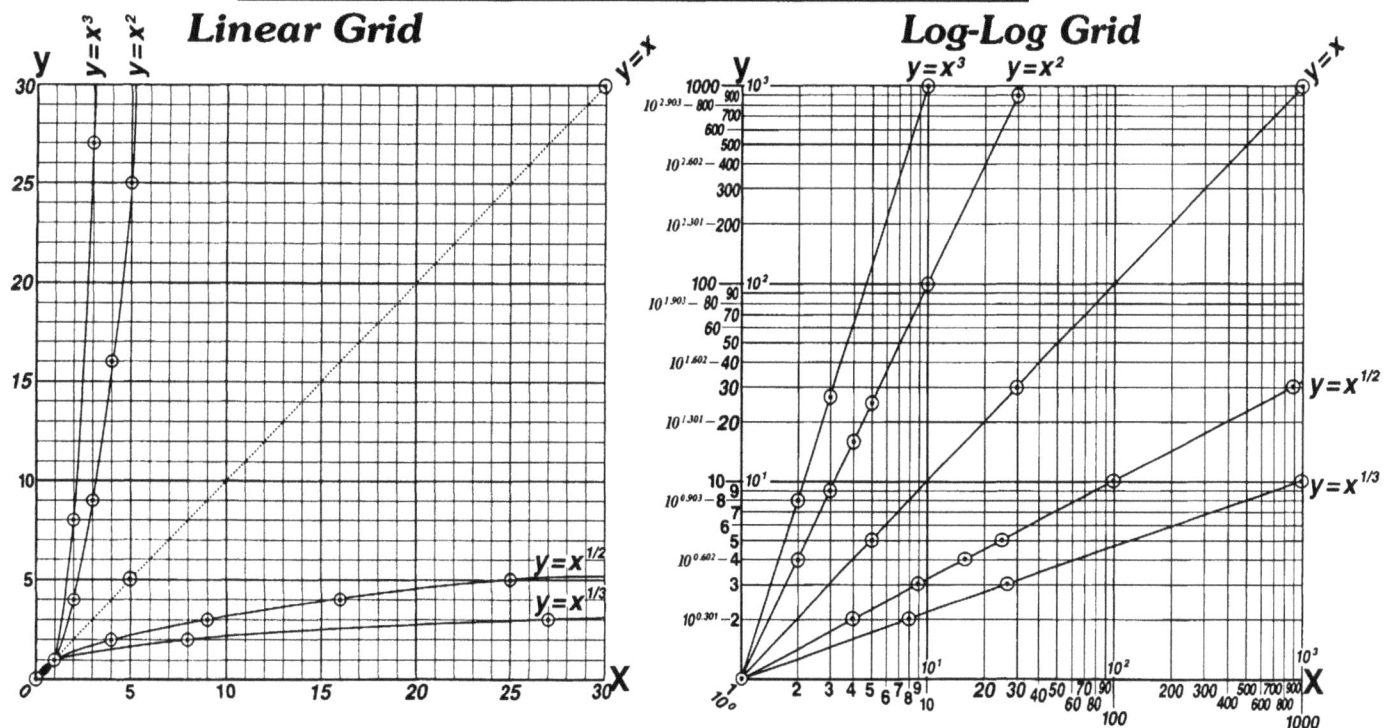

Linear Grid   *Log-Log Grid*

NOTES: **Activity E3**

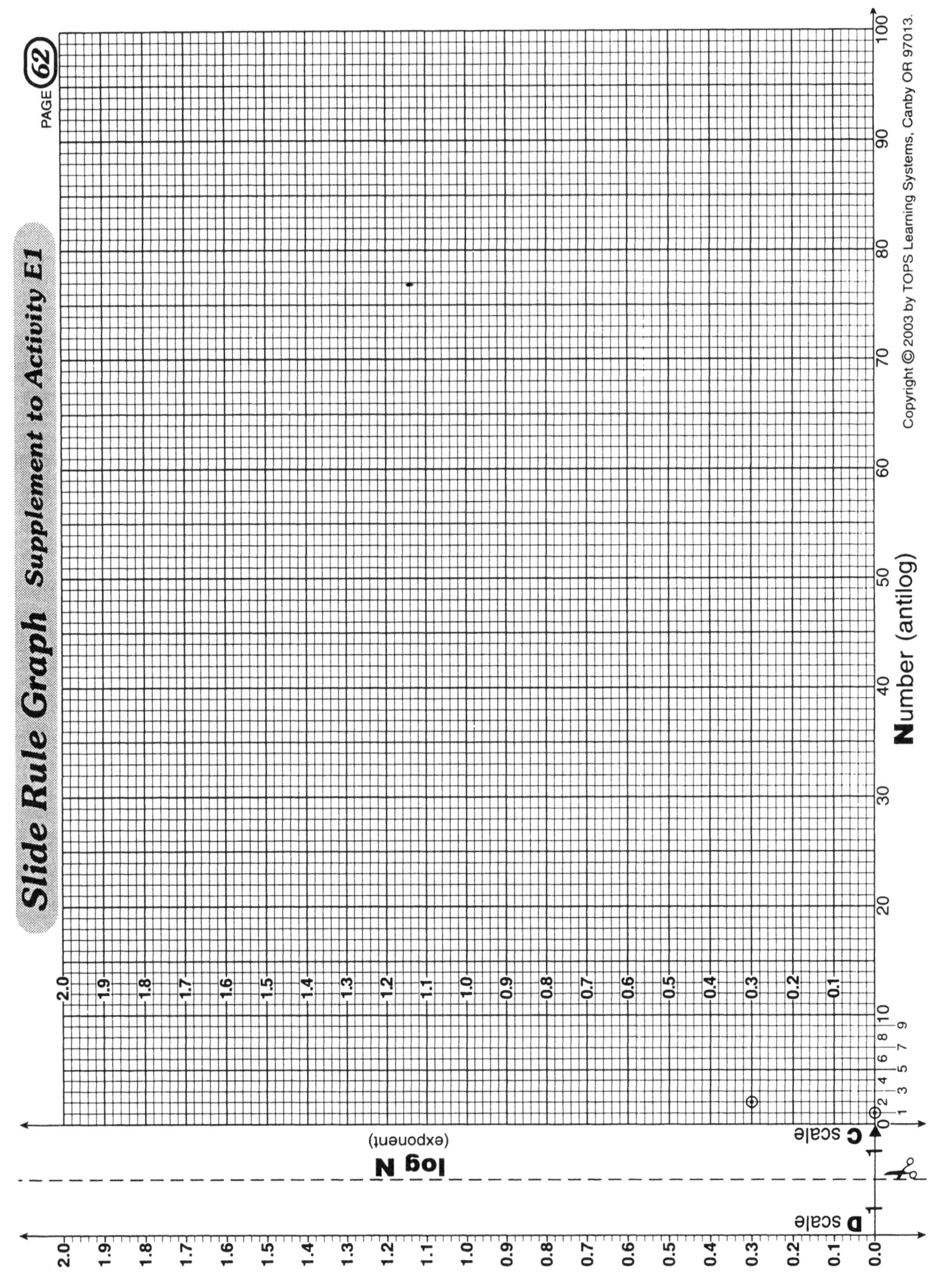

INVERSE FUNCTIONS
Supplement to Activity E2

Function	Ordered Pairs (x,y)								
y = x	(-2,) (0,) (2,) (5,) (10,)								
y = 2^x	(3,) (2,) (1,) (0,) (-1,) (-2,)								
y = 10^x	(1,) (.845,) (.602,) (.301,) (0,) (-0.155,) (-0.398,) (-0.699,) (-1,) (-2,)								
y = log$_2$ x	(8,) (4,) (2,) (1,) (1/2,) (1/4,)								
y = log x	(10,) (7,) (4,) (2,) (1,) (.7,) (.4,) (.2,) (.1,) (.01,)								

Note above row y = 10^x: −1.845, −1.602, −1.301

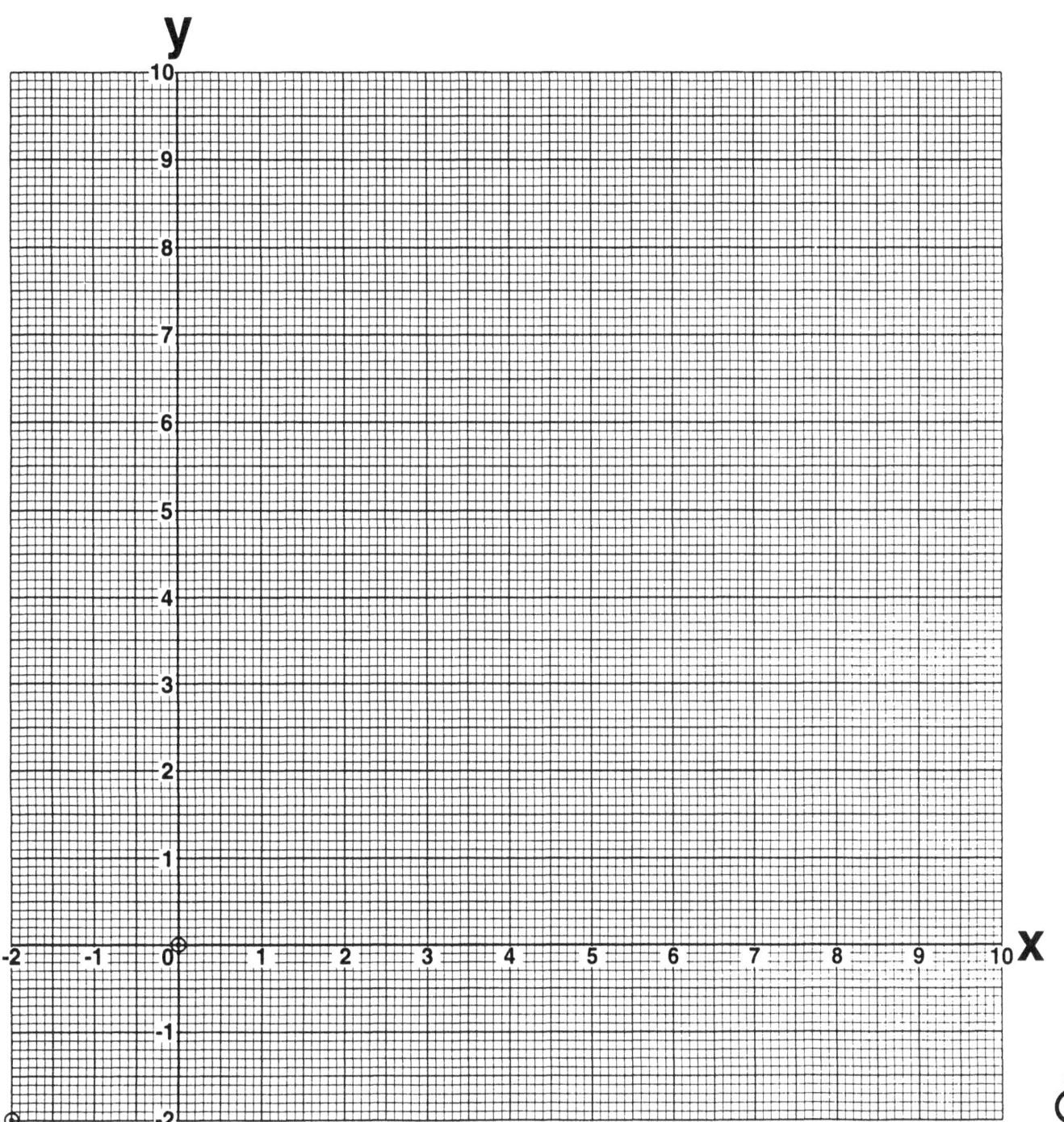

LINEAR/LOG GRIDS
Supplement to Activity E3

Log-Log Grid

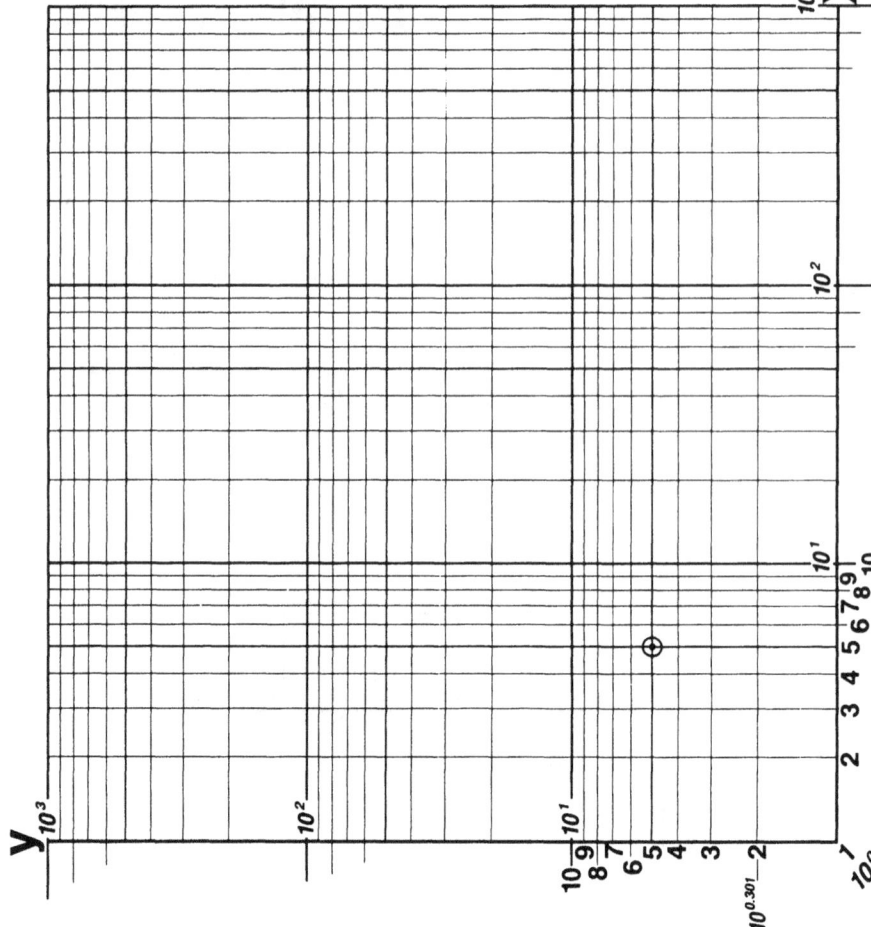

Linear Grid

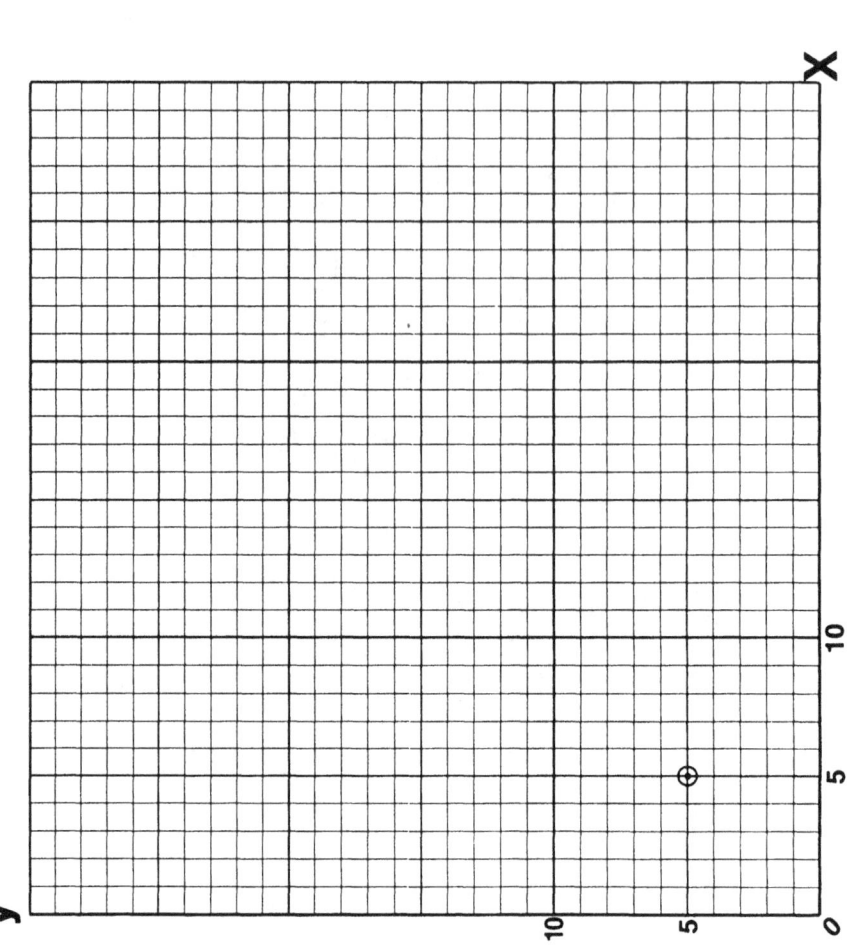

Functions	Ordered Pairs (x,y)				
$y = x$	(0,)	(1,)	(5,)	(30,)	(1000,)
$y = x^2$	(0,)	(1,)	(2,)	(3,)	(4,)	(5,)	(10,)	(30,)
$y = x^3$	(0,)	(1,)	(2,)	(3,)	(10,)
$y = x^{1/2}$	(0,)	(1,)	(4,)	(9,)	(16,)	(25,)	(100,)	(900,)
$y = x^{1/3}$	(0,)	(1,)	(8,)	(27,)	(1000,)

Feedback

If you enjoyed teaching TOPS please tell us so. Your praise motivates us to work hard. If you found an error or can suggest ways to improve this module, we need to hear about that too. Your criticism will help us improve our next new edition. Would you like information about our other publications? Ask us to send you our latest catalog free of charge.

For whatever reason, we'd love to hear from you. We include this self-mailer for your convenience.

Sincerely,

Ron & Peg

Ron and Peg Marson
author and illustrator

Your Message Here:

Module Title _____ Date _____

Name _____ School _____

Address _____

City _____ State _____ Zip _____

―――― FIRST FOLD ――――

―――― SECOND FOLD ――――

RETURN ADDRESS

_____
_____
_____

PLACE
STAMP
HERE

TOPS Learning Systems
342 S Plumas St
Willows, CA 95988

TAPE HERE

www.ingramcontent.com/pod-product-compliance
Lightning Source LLC
Chambersburg PA
CBHW081940170426
43202CB00018B/2963